PROFESSIONAL STUDIES IN ARCHITECTURE
A PRIMER

STEPHEN BROOKHOUSE

RIBA ##‖ **Publishing**

British Library Cataloguing in Publications Data
A catalogue record for this book is available from the British Library.

Commissioning Editor: James Thompson
Project Editor: Neil O'Regan
Designed by: Kneath Associates
Printed and bound by W. G. Baird

CONTENTS

ACKNOWLEDGEMENTS

I would like to thank Matthew Thompson for his editorial guidance, gentle persistence and constructive criticism and my family for their patience. I would also like to thank friends and colleagues for their valuable comments.

Special thanks should also go to Justin Nichols at Make Architects for the generous use of photographs from current projects and of the London office. Thanks also to Will Pryce, Terry Gamble at Denne and the London and Bath Estates.

DEDICATION

This book is dedicated to my family - and especially Betty and Susan Brookhouse

1
SETTING THE SCENE

INTRODUCTION

Design is the architect's core activity and we deploy our design skills to create buildings, structures and settings that people use and enjoy. At the end of the day it is the product that we stand back and critically evaluate or admire. This includes its physical properties – how it responded to the site and its context – how it performs in response to environmental conditions – heat, wind and rain. But what about the processes that took place that we cannot see that led to the physical product? How did the design respond to political, social and historical factors and Society's views of what it will accept – some of which may conflict with what the architect intended? These may have more influence on your design than the physical environment. How did a team of people come together to transform the design into a physical reality? How were they organised? How was the creative design process, which you know can be ill-defined and difficult to explain, managed in a way that satisfied the key stakeholders in the project – the client and the users? How did the project team manage the risky processes of construction? There is a product and there is a process but what characterises the knowledge, skills and professionalism of the key players – and the architect in particular – in this process? This book aims to explore and answer these questions and in doing so illuminate the processes that all contribute to successful architectural design and the construction of the built environment.

Professional studies in architecture, as a subject, often sits awkwardly with the core work in the design studio. The subject is frequently taught as sets of rules that are to be followed and a view of the architectural profession that bear little relation to your experience or your studio work. Although this book does not promise to fully integrate the subject areas with your design work it does aim to make the topics accessible in a way that is relevant to your work both in the studio and the early years in architectural practice. It does this in two ways: first, by explaining the context of each topic – and how seemingly impenetrable obstacles to building design have evolved due to social and political pressures as a set of invisible and changing constraints; and second, by acting as a filter for what is a very complex, detailed and evolving subject area. There is a trade-off between making the topic accessible and making it comprehensive. Therefore each chapter is intended as an introduction to the context and principles that govern each topic. Once you understand these you can then more confidently drill down to the detail covered in the main points of reference cited at the end of each chapter that practicing architects will routinely refer to.

The topics that comprise professional studies as a subject area have not been brought together before in one place nor been discussed and explained in this way. In explaining the context of each topic the aim is also to create links between subjects where relevant, rather than discuss them as separate 'stand-alone' subjects. For example, the concept of 'safety' in design, construction and building occupation has been a major concern of Society for centuries but has emerged as a fragmented set of different rules and regulations. The design of the urban and rural environment has developed in two very distinctive ways: how to stimulate and control development and how to preserve our heritage and quality of life.

To introduce the key themes of the book I want to take you on a short imaginary journey – a mini architectural tour. None of the buildings are particularly important or memorable but the journey raises questions that give a context to the topics discussed in the book. You can perhaps think of the book as an 'app' – a voice that gives many of the answers to the questions on the journey, that will help to also raise relevant questions giving appropriate answers to the unseen constraints that act on your design decisions and the processes and professionalism required to make your designs a reality.

HERE WE GO...

I am standing on a ridge in the Chilterns – part of a rough semi-circle of hills about 50 kilometres to the north of central London. I am surrounded by countryside – most of it is used for agriculture but some of it is set aside for recreation – rambling primarily. Looking along the ridge I can also see a monument to long-forgotten wars, The Boer Wars, fought in another hemisphere. Why do we remember it and does it have a significance other than remembering past losses?

THE BOER WARS WERE A CATALYST FOR SOCIAL CHANGE – ESPECIALLY HOUSING AND LAID THE FOUNDATIONS FOR MODERN TOWN PLANNING LEGISLATION.

The monument sits surrounded by land owned by the National Trust – an organisation that cares for and protects some of our heritage. Why is this 'open-access' land privately owned?

THE NATIONAL TRUST GREW OUT OF A PRESSURE GROUP IN THE NINETEENTH CENTURY IN RESPONSE TO THE INDUSTRIAL REVOLUTION. IT HAS MORE MEMBERS THAN ALL OUR POLITICAL PARTIES COMBINED.

Beyond the monument I can see the silhouette of a 'barrow', a large ancient burial mound, grazed by sheep. Built by our ancestors these massive earthworks are some of the first major man-made interventions in the landscape: its historical value is now largely forgotten and its purpose misunderstood. Is this protected too?

Looking the other way all I can see are fields and woodland dotted with small villages with large churches – their towers standing out. Nearby is a small ancient quarry, now pasture – but fenced in – with signs and interpretation panels telling us that it is now home to rare wild orchids. Is it also protected?

IT IS AN 'ANCIENT MONUMENT' – A RELIC OF OUR PAST, PRESERVED BY THE GOVERNMENT FOR THE NATION. THE ACT OF PROTECTION – ONE OF THE EARLIEST EXAMPLES IN THE LATE NINETEENTH CENTURY OF STATE INTERVENTION TO PRESERVE THE PAST – BUT CONVENTION AT THE TIME SAW THE BARROW AS AN OBJECT IN AN OPEN-AIR MUSEUM RATHER THAN AN ENVIRONMENT TO BE PRESERVED. IT TOOK MANY YEARS BEFORE THE 'ATMOSPHERE' OF THE PLACE BECAME SUFFICIENTLY VALUABLE TO PRESERVE.

YES, IT IS AN AREA OF 'SITE OF SPECIAL SCIENTIFIC INTEREST (SSSI)'. IT IS VALUED FOR ITS ECOLOGY AND HABITAT.

And why, when London is not far away by road or train is there little housing or other development?

The ridge borders an agricultural plain. Nearby, to the north, about 10 kilometres away I can see a small market town, Aylesbury. In the middle on a small hill sits a church, St Mary's – I can see its tower. Jumping in my car, (there is no public transport to speak of) I drop down to the plain and head towards the town.

YOU ARE STANDING IN THE 'METROPOLITAN GREEN BELT' – AN AREA AROUND LONDON AND OTHER MAJOR CITIES THAT OWE ITS ORIGINS TO CONCERN IN THE 1930S ABOUT UNCHECKED URBAN SPRAWL AND LED TO WIDESPREAD PROTECTION OF RURAL AREAS FROM DEVELOPMENT OF ANY KIND OTHER THAN AGRICULTURE.

As I get closer the road is bordered by non-descript semi-detached houses built in the 1930s. They are only one house deep with fields beyond.

As I get into the town I get snarled up in traffic. In the middle there is a ring road – impossibly close to the town centre – that cuts the town in two. It is a dual carriageway that forms a circle with roundabouts at key junctions and cuts through the pattern of streets and houses. Why was this built and why is it so close to the town?

THE RING ROAD WAS BUILT IN THE 1960S IN RESPONSE TO GREATER CAR OWNERSHIP AND THE SUBSEQUENT TRAFFIC CONGESTION. IT USED THE TEMPLATES SET OUT IN AN INFLUENTIAL BOOK PUBLISHED IN THE EARLY 1960S – 'TRAFFIC IN TOWNS' – THAT IMPORTED AN AMERICAN MODEL OF SEGREGATED CIRCULATION, TRAFFIC GRIDS AND NODES AIMED AT KEEPING TRAFFIC MOVING IN URBAN AREAS. BY SEPARATING PEDESTRIANS AND TRAFFIC ROAD SAFETY WOULD ALSO BE IMPROVED. THE NEW TRAFFIC ROUTES ALSO AIMED TO CREATE BOUNDARIES AND OASES THAT WOULD PRESERVE THE ATMOSPHERE AND VIBRANCY OF TOWN CENTRES. TOWNS AND CITIES JUMPED ON THE BANDWAGON – AND THE LEGACY CAN BE SEEN IN THE WAY THE URBAN LANDSCAPE HAS BEEN CARVED UP AND STREET PATTERNS DESTROYED IN THE HOPE OF KEEPING TRAFFIC MOVING.

Crossing the dual carriageway that cuts through the town I struggle to find a parking place in the multi-storey car park but having got there and back on foot I can walk up the High Street past the charity shops and estate agents windows. Where have all the 'real' shops gone?

THEY HAVE BEEN DISPLACED OR CLOSED DOWN AS A RESULT OF THE EDGE-OF-TOWN SHOPPING CENTRES WITH AMPLE PARKING – AN EXAMPLE OF HOW THE LOCAL GOVERNMENT PLANNING SYSTEM CAN FACILITATE GROWTH AND CHANGE BUT ALSO HAVE UNINTENDED CONSEQUENCES.

Climbing towards the church I enter an area of calm and quiet – the streets are too narrow to take too much traffic. Is this one of the oases envisaged by Traffic in Towns and has the traffic zoning worked?

YES, IT PROBABLY HAS WORKED – BUT LOOK CLOSER.

The scale and age of the housing around the church and its churchyard are very different. The streetscape is primarily made up of housing – with some offices in what were houses, a pub or two and a few small corner shops selling most things that you need. Although when you look closely there is considerable variety in style and scale. There is a sense of continuity and history – all the properties were built at least a hundred years ago and some must go back to the seventeenth century. Some of the houses and the rectory by the church in particular stand out as being well-designed. The area seems frozen in time. There are no neon street signs or modern buildings. This may be an oasis but why are there so few – if any – new buildings? It must be possible to place modern, contemporary buildings in this streetscape. ●

Walking back down into the commercial centre of the town I get to the market square – it is traffic-free, surrounded by a mix of pubs and mobile phone shops. The market square is not very large but at the far end is a huge, imposing building – the Crown Court. Why is this here, in a small market town? ●

THIS AREA OF THE TOWN IS PROTECTED BY LAW – IT IS A 'CONSERVATION AREA'. YOU CANNOT DEMOLISH, EXTEND, ALTER OR ADAPT ANY BUILDING WITHOUT PERMISSION FROM THE LOCAL AUTHORITY. THESE AREAS WERE CREATED AS A REACTION TO THE DAMAGE BEING DONE BY TRAFFIC SCHEMES AND NEW DEVELOPMENT IN THE 1960S.

WITHIN THE CONSERVATION AREA THERE ARE ALSO SOME BUILDINGS OF SPECIAL ARCHITECTURAL OR HISTORIC INTEREST. THESE ARE CLASSIFIED WITH DIFFERENT GRADES OF IMPORTANCE AND THEIR DETAILS ARE HELD ON A NATIONAL GOVERNMENT LIST, HENCE THE TERM 'LISTED'. THESE BUILDINGS ARE GIVEN GREATER PROTECTION AND YOU CANNOT ALTER THE BUILDING – INSIDE OR OUTSIDE – WITHOUT PERMISSION. IT IS A CRIMINAL OFFENCE. IN FACT THIS AREA HAS DEEP HISTORIC ROOTS – THE ROAD TO THE CHURCH HAS ROMAN ORIGINS (AKERMAN STREET), THE 18TH CENTURY CHURCH SITS ON A LATER ANGLO-SAXON SITE.

THE TOWN IS SMALL BUT IT IS THE ADMINISTRATIVE HUB OF THE COUNTY. THE LOCAL CROWN COURT HEARS CRIMINAL CASES AND PUNISHES OFFENDERS – SOME BEING SENT TO THE LOCAL PRISON ON THE OUTSKIRTS OF THE TOWN. THERE IS ALSO A COUNTY COURT IN THE TOWN WHICH DECIDES CIVIL CASES – DISPUTES ABOUT CONTRACTS, LAND DISPUTES, FOR EXAMPLE. THIS IS PART OF THE ENGLISH LEGAL SYSTEM: CRIMINAL COURTS WHICH HAND DOWN PUNISHMENTS AND CIVIL COURTS WHICH DECIDE DISPUTES. ORIGINALLY ALL CASES WOULD HAVE BEEN HEARD IN THE SAME COURTHOUSE.

BEHIND THE COURTHOUSE IS A GRAND TOWNHOUSE – THE 'JUDGES LODGING'. THIS IS A THROWBACK TO THE TIME WHEN COMMUNICATION WAS DIFFICULT. JUDGES TRAVELLED THE COUNTRY TO HEAR CASES IN DIFFERENT TOWNS AND CITIES. THE 'COMMON EXPERIENCE' OF DIFFERENT JUDGES WRITING DOWN AND COMPARING THEIR DECISIONS IS ONE OF THE FOUNDATIONS OF THE ENGLISH LEGAL SYSTEM.

Heading back towards the ring road there is a new theatre, the 'Waterside'. It takes its name from a branch of the Grand Union Canal that runs next to the site and which terminates in a small basin nearby where narrow boats were moored and turned. The narrow boats that moved raw materials, fuel and goods are long gone and even the pleasure boats have moved on. The theatre sits in what has become the commercial district of town and is surrounded by office blocks built in the 1970s and 1980s. This is a large, complex building – skilfully designed and well-built.

'How did the design evolve to its current form? Were the public given a say? Who made the key design decisions? What processes defined the size and shape? How was this complex building managed? Who were the key people in delivering the design and the construction?

THESE ARE DIFFICULT QUESTIONS ABOUT THE PROCESS OF DESIGN AND HOW IT IS MANAGED. IT INVOLVES COMPLEX RELATIONSHIPS AND A CLIENT WITH A VISION – IN THIS CASE THE LOCAL AUTHORITY – AND FUNDS – AGAIN FROM THE PUBLIC PURSE. A DESIGN TEAM OF DIFFERENT PROFESSIONALS, ARCHITECTS, ENGINEERS, ACOUSTIC DESIGN AND COST CONSULTANTS WORKED TOGETHER WITH THE CONTRACTOR'S TEAM AND HIS SPECIALISTS TO CREATE THE BUILDING. BOTH THE DESIGN AND THE CONSTRUCTION HAD TO BE CAREFULLY MANAGED TO ACHIEVE THE CLIENT'S OBJECTIVES AND THEIR BUDGET. THE PROCESS REQUIRED THE PARTICIPATION OF EXPERIENCED PROFESSIONALS – ALL WITH THEIR OWN SPECIALIST KNOWLEDGE AND SPECIFIC SET OF SKILLS WORKING TOGETHER RECOGNISING THE DIFFERENT CONTRIBUTION THAT EACH PROFESSIONAL BRINGS TO THE DESIGN AND CONSTRUCTION PROCESS. WHAT THE CLIENT WANTS WILL NEED TO BE ARTICULATED IN THE BRIEF AND THE PROCESS CLOSELY MANAGED.

THEATRES – AS PUBLIC BUILDINGS – HAVE TO COMPLY WITH MANY SETS OF REGULATIONS. THE BUILDING HAS TO BE SAFE FOR THE PUBLIC TO USE – WITH ADEQUATE FIRE ESCAPES FOR EXAMPLE, BEEN BUILT SAFELY AND BE SAFE TO MAINTAIN IN THE FUTURE. ALTHOUGH YOU SEE THE END PRODUCT THIS HAS ONLY BEEN ACHIEVED BY A COMPLEX SET OF PROCESSES INVOLVING, DESIGN, REGULATION AND CONSTRUCTION THAT REQUIRE AN ORGANISED, SKILLED TEAM WHO MAKE IT HAPPEN. ALTHOUGH YOU CAN SEE AND FEEL THE BUILDING AND ENJOY USING IT, THE DESIGN VERY RARELY GIVES A DIRECT INSIGHT INTO THESE COMPLEX PROCESSES.

Perhaps this is a good point at which to stop and enjoy a show. This is a journey that anyone can make in any town or city. You can ask yourself similar questions and hopefully, having read this book, you will find the answers too.

This book discusses the wider, and unseen non-physical constraints that affect the built environment, the professional qualities of the specialists who make projects happen and the different ways that the design and construction phases of design projects are organised and why. Throughout the book I have deliberately placed each topic within its historical and/or global context. This sometimes involves making links across seemingly separate areas – pollution and sustainability for example, and placing topics such as English Law and urban planning legislation within a continuum.

It is important to understand the importance and the nature of this continuum as the professional studies landscape is constantly changing – due to socio-political, global and physical forces. For example, while the science of global warming may continue to be questioned by a diminishing minority of dissenters, the effects of population growth, economic growth and poverty are clear – and these are an integral part of our approach to global sustainability which results in national and local initiatives. Because professional studies in architecture are subject to these forces the scope and detail of each topic area can change rapidly. An example is urban design policy. The chapter on planning has been changed three times to reflect policy changes in the short period it was being drafted. The Building Regulations for England and Wales are under continuous review and due for major changes. The legality of our health and safety rules is being questioned as a result of investigations by the European Community. Government tinkers with its green economic policies – constantly changing incentive schemes and costs to try and make them affordable and sustainable while promoting economic growth that itself will consume scarce resources.

Architects need to remain competent and need to maintain an understanding of where these rules and regulations come from as well as to remain knowledgeable of where they are going. Through this better understanding you will also be able, in time, to contribute in an informed way and influence the future direction of the architectural profession and its role within the urban environment and society.

2
THE DESIGNER AS PROFESSIONAL – WHAT IT MEANS TO BE AN ARCHITECT

Many factors help to shape the built environment. Each building has its own complexities that concern location, site, context, client brief and are resolved in both the design and its construction. As well as the designer's creative imagination, a high level of knowledge, skill and competence underpin the design. Not all architectural products are exceptional, ground-breaking or innovative but they all serve a function and respond to a need. Their building is carried out by a skilled team of design and construction professionals who are all stakeholders in the process as well as the shape and appearance of the product.

The purpose of this chapter is to investigate the idea of the architect as a design professional. What characterises a professional as different from other occupations? How do professional organisations shape and perpetuate their members' status? How has the view of professionals in society changed and how will it develop in the future?

You're reading this because you want to be an Architect with a capital 'A', a professional building designer. You probably feel instinctively that being an architect is different from working as a 'job runner' or an 'architectural assistant' in a practice. But is it possible to define this difference?

Is part of the definition to do with what an architect knows – our silo of knowledge? Believe it or not, you have already built a body of knowledge and certain skills and begun to apply and develop your competence to contribute to the built environment. This knowledge set is different from that being learnt by other students at university and, if you have experience of architectural practice, other non-architect professionals in the construction industry. But what is it about this knowledge that makes it 'professional'? After all, others – non-architects – can and do carry out the same kinds of jobs competently.

FIGURE 2.1

The word 'professional' is used in many ways. Is the professionalism of an architect different from that of a professional footballer or professional boxer – or a professional poker player? Does a professional hypnotherapist have the same status as clinical psychologist or a doctor? We know that there are differences, but what are they? The discussion below should help you to understand the differences and also the special qualities of the design professional.

DEFINING THE PROFESSION

Sociologists have debated the term 'profession' without coming to a firm conclusion. For example, one view is that 'there is no logical basis for distinguishing between so-called professions and other occupations.'[1] Others observe that the term is a lay or folk term and that the public in general assess the 'traits' of occupations with different levels of precision. As customers, patients or clients, we are very aware of the performance of the professions and come to a conclusion based on our experience. The professions themselves and the state make specific decisions which determine the standing of a profession. Conventionally, most authorities agree that the professions are defined by four characteristics that distinguish them from other occupations.

They must:

1.   Have their own distinct body of knowledge.

2.   Erect barriers to entry to maintain standards.

3.   Serve the public interest.

4.   Enjoy mutual recognition from other professionals.

A DISTINCT BODY OF KNOWLEDGE

Architects have their own body of knowledge (often described as a 'silo' of knowledge) that is distinct from that of other members of the design and construction team. It allows you to communicate with other members of the profession in a way that is almost impenetrable to others. For example, for an architecture student, the iconic twentieth-century architect Mies van de Rohe's 'less is more' is far more than just a quotation. It is shorthand for a particular design approach and a whole body of work. Other examples are the 'charette' as a way of working, or the 'crit' as a method of evaluating work. In time, these common points of reference and experiences accrete to form the walls of the silo. They reinforce the sense of a professional grouping. Even the distinct working methods in schools of architecture, together with the length of professional training, help to reinforce the idea that architectural training is different.

Professional bodies naturally articulate the things that distinguish them from other professionals. This leads to a core set of competencies which are set down and reviewed on a regular basis. For architecture students, these are stated in a set of Criteria, 'The General Criteria for Parts 1 and 2 and the Professional Criteria for Part 3.'

Students cover the Criteria through a mixture of formal learning and practical experience. Schools of architecture are tasked with delivering these Criteria through their courses, which are in turn regulated by the profession through the professional membership organisation, the Royal Institute of British Architects, (RIBA), and the statutory regulator, the Architects Registration Board (ARB). (The history, objectives and function of these two organisations are discussed in more detail below.) Architectural practice nurtures students during compulsory periods of practical training, supervising and mentoring their professional development. In this way an 'academy' in the shape of schools of architecture, the profession in practice and the professional bodies work together to define and refine professional competencies.

There are risks inherent in this process of developing the body of knowledge and professional competencies that on balance, the architectural profession appears prepared to accept.

First, by defining the limits of professional competence, the role and influence of the profession is also limited. Expand the silo and you risk diluting existing expertise or encroaching on the territory of other professionals.

Second, this process takes time – indeed many years – and during this time ideas and the industry can move on. The architectural profession faces a particular set of pressures because it sits within and possibly to the side of the construction industry as a whole. There are also much bigger professions within the industry that have a louder voice.

Economists describe the development of specialisms as inevitable and a manifestation of the 'division of labour', a principle identified by the father of economics, Adam Smith, in his 'Wealth of Nations'. He argued that as industrial processes become more complex, we respond by breaking the production process into smaller parts and perform those smaller tasks more competently and with greater efficiency. (Smith famously described the process of making pins.) Certainly, the profession has identified that the role of the architect has become more specialised as the construction industry has become more complex and industrialised. The architectural profession is perhaps more susceptible to these pressures of industrialisation than other professions. In medicine or the law, for example, the professional skills exist in a highly regulated and controlled environment where they are perhaps in a better position to control the effects of industrialisation and technology.

MAINTAIN STANDARDS BY ERECTING BARRIERS

As a pre-condition of creating a profession, the idea of maintaining standards by erecting barriers is obviously closely linked to the body of knowledge. First, the time and aptitude to qualify as an architect is a barrier in itself – not everyone has what it takes. Also, the design process is difficult to articulate to non-practitioners, making it relatively exclusive.

However, having a unique set of references and a modus operandi are not enough automatically to create a set of professional standards. Instead, the profession sets levels of knowledge and competence that students must meet or exceed, and which practitioners must maintain. The barriers and standards are set by the schools of architecture, practitioners and the professional organisations. The schools are tasked with maintaining these standards, all monitored by the profession by means of periodic reviews.

The barriers to entry in architecture are high and difficult to scale. As the skills and knowledge required for Parts 1 and 2 are normally delivered in a school of architecture, they follow normal undergraduate and graduate academic and assessment procedures. In effect, the standards are set by the schools on behalf of the profession. Part 3, the Professional Examination, follows a similar path but at this point – the point of entry to the profession – it sets out specific procedures for the examination where professional architects with suitable experience examine candidates and the schools have to follow these procedures or risk losing recognition from the profession. As this recognition is effectively a licence to deliver architectural courses, losing professional recognition would have a serious adverse impact on a school's survival. The profession, in turn, would also lose its ability to maintain barriers and standards. Therefore there is a level of self-interest that sustains the keepers of the standards until those standards lose their professional value and cease to be relevant. The danger then is that professional education leads to an unwillingness or inability to step beyond the boundaries that defines the architect's silo of knowledge, a risk that has been termed 'trained incapacity'.

SERVING THE PUBLIC INTEREST

All professions try to distinguish themselves from the non-professions by claiming to promote the interests of the public and society at large before the interests of their membership. Historically, they have maintained their privileged status and recognition in society by avoiding any political allegiance and by not pursuing commercial self-interest. Typically, these principles are framed in a code of ethics – a set of moral principles.[2] Hence, lawyers proclaim that their first duty is to the law itself. Medical physicians take the 'Hippocratic oath', promising to treat all people fairly and to seek to preserve life.[3]

Architects have an especial obligation to society. Even though their designs are commissioned by clients to whom the architect will have a set of professional obligations, their buildings are in the public realm of the built environment and thus impinge upon society at large. The inherent permanence of buildings makes this effect lasting and disproportionate to original intellectual effort. It also makes a contribution to the body of knowledge and may be the focus of debate and architectural theory.

This moral perspective and wider impact is expressed by Alain de Botton. Talking about Herzog and de Meuron's conversion of the Bankside power station into the Tate Modern art gallery in London, he said:

'All works of design and architecture, from a parliament to a fork or cup, talk to us about the kind of life that would most appropriately unfold within and around them. They tell us of certain moods that they seek to encourage and sustain in their owners. While keeping us warm and helping us in mechanical ways, they simultaneously hold out an invitation for us to be specific sorts of people. They speak of particular visions of happiness. Hence to describe Tate Modern as beautiful suggests more than a mere aesthetic fondness; it implies an attraction to the particular way of life this structure is promoting through its roof, door handles, window frames, staircase and furnishings. A feeling of beauty is a sign that we have come upon a material articulation of certain ideas of a good life.'[4]

At a functional level, defending the public interest requires both setting and promoting the standards of competence one can expect from an architect. The theory is that this will ensure that those who hold the professional title will practice competently. This functional requirement is expressed in the ARB's mission to protect the consumer and safeguard the reputation of architects.

Protecting the public interest is one of the key attributes that distinguishes the professional from other highly skilled occupations. Both the professional footballer and poker player will have a code of ethics and behaviour but may not meet the wider requirements of defending the public interest.

ENJOY MUTUAL RECOGNITION FROM OTHER PROFESSIONALS

One of the starting points of this chapter was that the design and construction of the built environment is a team endeavour. At a project level, it is important that the specialist skills, knowledge and competence of each professional team member are recognised by the other members of the team. In short, the team will only function effectively if it has the right skills. Being a member of a recognised profession is a signal to other team members of a set level of competence. Therefore it is important that each profession communicates its expertise as widely as possible and seeks to promote its standards. Within the project team there are many specialists: engineers, cost consultants, designers, project mangers and construction managers. As projects become more complex, more specialists are required to manage the process and design the components that come together to form a building.

The mutual recognition from other professionals is also reinforced by the profile of individual professions in the eyes of the public, business and government. They are considered strong or weak depending on the level of influence they can exert on decision-makers. In this way the professions are also seen as interest groups and the line between promoting their higher order objectives in general ('architecture' or 'medicine', for example) and the interests of its members ('architects' or 'doctors') is a fine one. However, the professions expend considerable efforts in promoting themselves not only to other professionals but also government and the public.

'ARCHITECT' – A PROFESSIONAL TITLE OR A PROFESSIONAL ROLE?

In the United Kingdom, the *role* of the architect in the construction industry or society generally is not protected. Anyone can set themselves up and do what an architect does. However, they cannot use the *title* 'architect'.

This is very different from some other states in the European Union, where certain functions and tasks are heavily regulated. For example, in the UK anyone can submit a planning application. In many other EU states, this process (or its equivalent) has to be carried out by an architect registered in that state. It is also different from the conditions that apply to other established professions in the UK, such as doctors, lawyers and accountants, who have a protected role in certain specific functions regulated by the state. Therefore society and the state are not consistent in their approach to the roles of the different professions. This presents a challenge to the professional organisations that are responsible for and represent architects.

PROFESSIONAL ORGANISATIONS IN ARCHITECTURE

The architectural profession is represented by two different organisations – the RIBA (Royal Institute of British Architects) and the ARB (Architects Registration Board). The RIBA is commonly known as the professional body and the ARB is known as the regulatory body. In other EU states there is generally only one professional body.

This section explains how the organisations came into existence, where they obtain their powers, what or who controls them and their key functions.[5]

THE RIBA

'To advance architecture by demonstrating benefit to society and promoting excellence in the profession.'[6]

The RIBA is a conventional professional body: a learned society with a membership of over 40,500 members worldwide. It declares its mission as the promotion of architecture as well as upholding the standards of the profession.

It was granted its first Royal Charter in 1837 with its origins as a learned society. Its formation occurred at a time of great political, social and economic change in Britain. The industrial revolution was underway with its consequent radical changes in the composition of British society. There were also fundamental political changes taking place, culminating in the Great Reform Act of 1834. The growth of the architectural profession and its recognition could be seen as a result of industrialisation, a principle identified previously by the great economist Adam Smith.

These changes were widespread. For example, the term 'scientist' was only coined four years earlier in 1833. Technical drawing had been introduced from France by Isambard Kingdom Brunel's father during the Napoleonic Wars. Engineering was a growth profession. Buildings and building types were becoming more complex, with the scale and size of public buildings increasing. The political and social elite were better educated and architecture as a profession became seen as a separate discipline from the building process that had its own historical and typically Classical rather than vernacular precedents. All of this and more led to a growth in demand for professional specialism generally, and as a result the discipline of architecture benefitted.

The RIBA is also governed by a set of bye-laws approved by the Privy Council,[7] which provides it with its operational framework.

Its vision is to be the 'champion for architecture and for a better built environment'.

The RIBA's strategy has five objectives:

1. Demonstrate the benefits of good architecture.

2. Promote and enhance the benefits.

3. Facilitate the delivery of good architecture.

4. Provide high-quality support services.

5. Develop the capability to deliver the strategy.[8]

As you would expect, it meets the four characteristics of a professional body.

- It has a prominent role in setting educational standards through the Criteria for Parts 1, 2 and 3 (held in common with the ARB) and working with schools of architecture.

- It has barriers to entry, where candidates for chartered membership must meet all the Criteria and have completed a minimum of twenty-four months recorded and monitored practical experience in a recognised professional setting.

- It serves the public interest – made explicit in its mission statement – and states its ethical position through its Code of Conduct, which it promotes to the public and which chartered members must follow. A disciplinary function serves to protect standards and conduct.

- It seeks mutual recognition from other professionals and works with numerous specialist professional interest groups in the construction industry, and promotes its vision to society. Recognition is also implicit in its chartered status and the regulation by government evidenced in the involvement of the Privy Council.

The RIBA also promotes itself across the globe. It does this by liaising with similar bodies worldwide, organising and attending international conferences. Equally, it promotes standards of architectural education through the recognition of foreign schools of architecture that adopt and follow its Criteria for Parts 1 and 2 and set equivalent standards.

As a membership organisation, the RIBA has an elected President and a Council of sixty members, the large majority of whom are chartered architects. Detailed responsibility for running the Institute is passed down to a Group Board, which directs the overall business of the Institute. The RIBA is led by a Chief Executive with a staff of approximately 200. Through its subsidiary companies, the RIBA deals with the promotion of architecture, professional education, specialist services and guidance to members and clients under a number of brands, such as RIBA Publishing, NBS, and RIBA Appointments.

Complaints are dealt with by a Professional Standards Department, which supports a Disciplinary Committee made up of members. Hearings are held *in camera* and the decisions are published in the RIBA Journal.

The RIBA has also been recognised as a 'Business Superbrand', offering quality, reliability and distinction in its field.

THE ARB

'Protecting the consumer – supporting architects through regulation.'[9]

The ARB was established by Parliament in 1997 to regulate the architectural profession in the UK. Its duties are contained in the 1997 Architects Act. In 2010, there were approximately 30,000 architects on the register.

The ARB's duties cover five main areas:

1. Prescribing – or 'recognising' the qualifications needed to become an architect.

2. Keeping the UK Register of Architects.

3. Ensuring that architects meet its standards for conduct and practice.

4. Investigating complaints about an architect's conduct or competence.

5. Making sure that only people on its register offer their services as an architect.[10]

'We are an independent, public interest body and our work in regulating architects ensures that good standards within the profession are consistently maintained for the benefit of the public and architects alike.'[11]

As a statutory body, the ARB is also the vehicle for enacting relevant European legislation. For example, it is the designated Competent Authority for providing recognition of qualifications from other EU states and for admitting suitably qualified applicants to the register under the *Mutual Recognition of Professional Qualifications Directive* (2005). (For more on how the UK parliament must incorporate European initiatives, see Chapter 5.)

Its objectives broadly align with those of the RIBA. It shares the same Criteria for Parts 1, 2 and 3 with the RIBA and sets its own barriers for entry for individuals educated outside the EU who want to register in the UK. It has a professional code of conduct – 'The Architects Code: Standards of Conduct and Practice'[12] – and disciplines registered architects who do not follow the standards of professional competence or behaviour set out in the Code. In addition, it has the role of protecting the use of the title 'Architect' and individuals or organisations that pass themselves off as such can be prosecuted. It shares mutual recognition with other statutory and professional bodies in the EU and the UK. Through these linked activities it defends the public interest and as a statutory body it must also be independent of political and commercial pressures and interests.

Its public interest mission is further reinforced by its composition. The ARB has a fifteen-member controlling Board with seven architect members elected by the profession itself and eight 'lay' members appointed by the Privy Council. All members of the Board serve for a fixed term and then stand down. The lay members are always in the majority. In 2010, for example, the Chair was a trained industrial design engineer and the Vice-chair was a registered architect.

The ARB appoints a Chief Executive who leads a small staff with three main departments: Qualifications, Registration, and Professional Standards. Disciplinary matters are dealt with by a Professional Conduct Committee made up of publicly-appointed registered architects and lay members. Hearings are held in public and the decisions are published. As part of its work, ARB also publishes information for the general public and guidance for registered architects on subjects such as professional indemnity insurance.

THE CODES OF CONDUCT

Both organisations use their Codes as tangible ways of showing the public interest function of the professions. They set higher standards of professional conduct and competence than is expected in a normal commercial relationship between customer and provider. All registered architects must follow the ARB's Code and Standards. If you are also an RIBA Chartered members, you must follow the RIBA's Code of Conduct as well.

ARCHITECTS AND ARCHITECTURAL PRACTICE

The key practical difference that derives from the different mandates of the ARB and RIBA concerns architectural practice itself.

The ARB's main focus is to regulate the title 'Architect', which applies to individual architects. Their role is limited to giving advice to consumers and individual architects, which in itself is very useful, but does not and cannot address wider concerns of architectural practice.

On the other hand, the RIBA *does* support architectural practice – how groups of architects come together, work together and work with other construction professionals, clients and users.[13] It provides services for the public, individual members, and promotes practices through its directories and website as well as holding and publishing details of individual members. It celebrates the architectural excellence of practice through its awards. For example, the RIBA awards generally are given for built work. These are seldom the work of individual architects but represent a group enterprise and a partnership between all stakeholders in a project. The vital role of those who commission projects is also recognised by an RIBA award for clients. An example of the recognition of the group nature of architectural projects is the RIBA award given for the Accordia project in Cambridge. This was the work of a number of different architectural practices working to a common brief on a single site.

A CRITICAL FRAMEWORK FOR UNDERSTANDING THE PROFESSIONS

The profession faces change in the future on a number of fronts: challenges to the status of its members, the role of government, the erosion of some of its functions with the emergence of co-professionals, and the value of a multi-disciplinary approach to construction projects.

The critical framework below is intended to give you some models to help you interpret the arguments for and against changes that have taken place. It will also help you interpret and map your own professional experience.

The different views on professionalisation over the last seventy years have been grouped as follows:

FUNCTIONALIST

The functionalist approach is the one that has generally been discussed so far.
It concentrates on functional 'traits', including service, altruism and professional
ethics ('the public interest') and the division of labour that leads to specialisation.
Essentially, it follows the idea that the professions are identified by what they do.
The threat to the professions is seen as coming from the rise of bureaucracy and
management becoming the dominant force in modern society.

MONOPOLY OR 'POWER' APPROACH

The monopoly or 'power' approach is concerned with the power of the
professions first to determine the boundaries of their own working practices and
then to assert their autonomy to prevent outside interference and supervision.
In order to achieve monopoly, i.e. an exclusive licence to practice, an occupation
must have a special relationship with the state to make a 'regulative bargain'.
How this is done will depend on the political culture but traditionally the trade-
off for monopoly has been an independence from commercial influence – a
professional disinterest. The trouble is, they sometimes must compete in the
marketplace with other occupations who can provide similar or complementary
services – the effort by the professions to defend, maintain and improve their
position.

In the UK the critical spotlight has focused on the relations between the
consumer (increasingly championed in recent years) and the producer of
professional services, and the worry that the producer controls this relationship.

CULTURAL

The cultural model concentrates on two things: the profession-specific values
and norms, which create the architects' collective silo of knowledge; and the way
architects promote their ethical standards and altruism.

THE PROFESSIONAL PROJECT

The work of the sociologist MS Larson is seen as central to the development
of theories about the professions in the late-twentieth century. Her key theme,
described as the 'professional project',[14] draws on the ideas of social stratification
in society and the professions' influence on the economic and social order so
as to capitalise on the value of the professions' specialist knowledge. Larson
summarises her approach as follows:

'Professionalization is thus an attempt to translate one order of scarce resources – special knowledge and skills – into another – social and economic rewards. To maintain scarcity implies a tendency to monopoly: monopoly of expertise in the market, monopoly of status in the system of stratification.'[15]

The following diagram offers a working model of the professions based on the ideas discussed. The term 'social closure' can be seen as a measure of both the monopoly and control of professional knowledge and services, state regulation and a high status in society. Strong professions will contain all elements whereas weaker ones will only share some and exert less control over others.

FIGURE 2.2

The Professional Project

THE PROFESSIONAL PROJECT

THE ECONOMIC ORDER

THE SOCIAL ORDER

Legal monopoly of knowledge-based services

High status and respectability

MONOPOLY OF KNOWLEDGE

The state needs services grants monopoly, achieves regulation

Culture specific values and norms

SUCCESSFUL OUTCOME

'SOCIAL CLOSURE'[16]

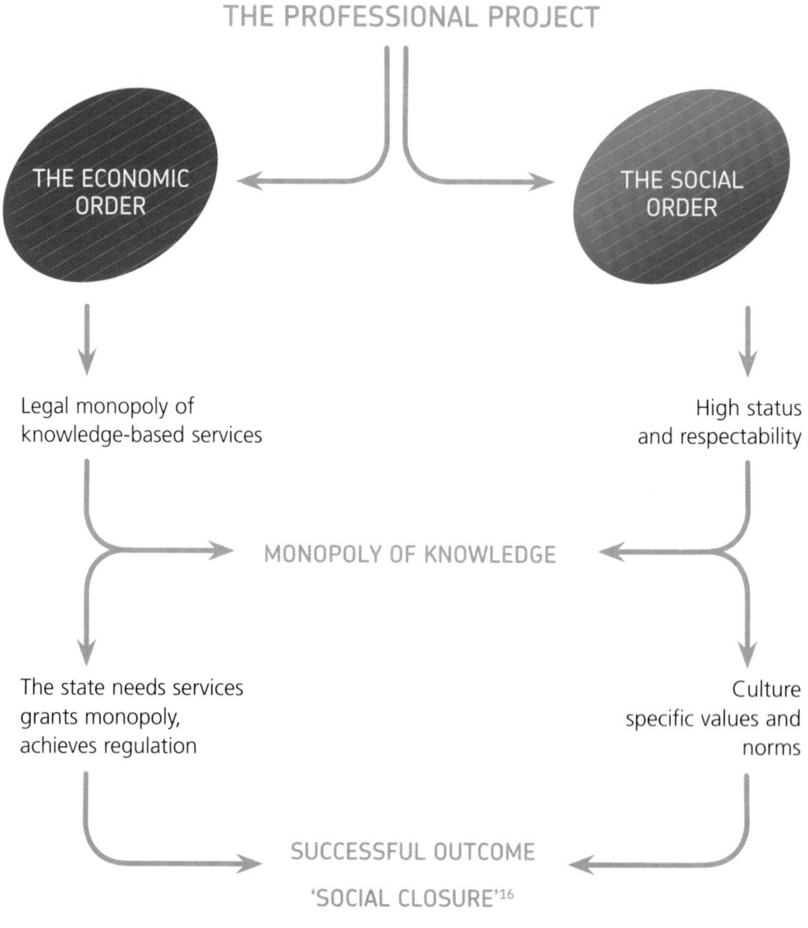

You can use this model to interpret the architectural profession. On the economic order side of the diagram, architects do not have a monopoly of knowledge-based architectural services. The role of the state is less evident in architecture than in other professions because the renewal of the government estate has gradually transferred to the private sector. In this context, regulation of architectural services can be seen as essential for the future survival of the profession. This is the role of the ARB, but statutory regulation only relates to the title not the role.

On the social order side of the diagram, architects have a high profile in part due to the public profile of their work – the buildings themselves. (Contrast this with the difficulty of communicating the cultural values of accountancy or the law.) This also helps to maintain architects' social status. The public interest and ethical position of architecture working for the public good are also strong.

As there is not a monopoly of knowledge or services, there is a lack of 'social closure' – in effect, a lack of control. In this analysis, architects are in a weaker position than, for example, doctors and lawyers. These professions are more successful at controlling the exclusivity of their services and exert considerable influence over the state by making their services indispensible.

THE CHANGING ROLE OF THE PROFESSIONAL IN SOCIETY

You can also use the professional project model to interpret the effect of wider changes on the professions and the architectural profession itself. The model is not comprehensive though and suffers from a number of drawbacks. First, in a general sense, when the social or economic order changes due to structural or attitudinal changes in society, or major economic events such as the recession of 2007–2009, the professions must respond or change too. Secondly, and more importantly, the context for the delivery of architectural services is very different from that of medicine (with essentially one client – the state) and the law (which provides a number of essential but generally isolated services). Unlike most other professions, architects work within a much larger industry. When the industry shifts, architects feel it.

All professionals are defined by the work they do – the functional model. Architecture and an interest in contributing to the built environment are the things that drive architects. These interests – attending to the public attitude to architecture and the relationships with clients and users – rightly occupy the architects' time. They have no day-to-day imperative to influence the state as well.

The **social order** has changed radically since the professions were first established in the nineteenth century. Since the 1960s, the growth of higher education, a dismantling of the historic class system and the growth of the affluent society have all allowed more social mobility. Professional services are perceived as being of higher value than manufacturing, which, as well as being good for the professions, is paradoxically also a threat. Information technology has freed up knowledge and made it increasingly difficult to control.

All this has meant that the perceived higher value of knowledge-based services has been balanced by the erosion of the status of the professions. The consumer has emerged as a powerful voice and is now far more ready to challenge the status of the professional and question the value of the professional services he or she receives. Consumers question the public interest aspects of the professions, their ethical stance, their altruism and their position of independence from commercial pressures.

Architects have arguably been affected by these changes more than other professions. People are much more likely to question the quality of the built environment and have a greater voice through the democratic nature of the planning system in the UK. The growth of consumerism also means that the consumer now has greater opportunity to dispense with the services of an architect altogether. The status of the architect as a designer, though, remains relatively robust.

The **economic order** has also changed. The state reflects society and is less likely to defend the legal monopoly of the professions. The current trend is for the state to take the regulatory function away from the professions and place it in the hands of an independent body. The way the architectural profession has evolved aligns with this trend.

The professions have not helped themselves. Lawyers and accountants who have legal and fiduciary duties as advisors and scrutineers in the running of large companies (in effect, the state's 'public conscience') have been found wanting. They have been at the centre of the corporate fraud that led to the collapse of major companies such as the international energy company Enron (described by *Fortune* magazine as 'The world's most Innovative company' for six years running and employing 22,000 people worldwide at its peak) in 2001 and the IT company Worldcom 2002.[17] The public rightly believed that the ethical standards and the independence from commercial pressures that secured the professions' status had not been maintained. The state's response has been to move towards greater regulation.

In this analysis, the architectural profession can be seen as relatively strong on the social order factors but comparatively weak on the economic order ones. Architects may have a higher social status than some professionals but suffer financially because control over their role is relatively weak.

RECENT PRACTICE

The architectural profession has attempted to reinvigorate itself in recent years in a number of ways. As long ago as the early 1960s, the RIBA carried out a study of the profession – 'The Architect and his Office'[18] – which highlighted the need to improve the management of the services architects provide to clients. In the early 1990s, the RIBA published the 'Strategic Review of the Profession' which again identified weaknesses in formal management processes, the relationship between architects and their clients, briefing and their role in multi-disciplinary teams.[19] The profession in mainstream practice has responded to these challenges and made major changes in the way their services are performed.

The profession's ethical codes have changed to respond to changes in attitudes and commercial pressures. As recently as the early 1980s, the RIBA Code of Conduct continued to reflect the concept of the professional's position of independence from commercial interests. For example:

1. architects in practice, in common with other professionals, could not operate as companies. They had to remain as professional individuals working in partnership with others rather than in a separate legal entity,

2. architects could not act as contractors or join contracting organisations,

3. architects could not advertise their services.

Eric Lyons, a relatively well-known and respected architect in the 1960s, had to give up his professional membership to pursue his interest in system-built housing when he formed Span, a company that designed and built relatively low-cost housing. Attitudes changed and Lyons went on to become President of the RIBA in the late 1970s. (Span housing, always respected by the profession, later became the subject of a book on the subject published by RIBA Publishing.) These three conditions of the Code of Conduct were removed in the early 1980s.

FUTURE PRACTICE

'Society no longer respects tradition for its own sake. Any profession must continually demonstrate its value in the public interest and must be prepared to accept scrutiny on a hitherto unprecedented scale.' Stephen Hockman QC[20]

Future architectural practice is what we do tomorrow and not just some distant future. For example, one of the biggest upheavals now results from major changes in the way buildings are procured, commissioned and built. It is therefore very important to consider how the architectural profession adapts. Architects will have to respond to the twin pressures of value in the public interest and increased scrutiny. At the same time, it will have to do this against a backdrop of the trend towards greater specialisation, continual changes in procurement methods, shifting public opinion about the professions in general, and an increasingly vocal consumer as stakeholder, client and user: all without the protection of role and function architects need.

The options facing waning professions range from 'diffusion' to, at the opposite extreme, 'closure'. In the first, the professions expand into new industries, diffusing their expertise. In the second, they retrench and erect higher barriers until membership dwindles.[21]

The construction industry has tried to reduce fragmentation and improve efficiency, encouraging teamwork and the formation of long term partnerships between clients, designer professionals and contractors. This has led to changing relationships between clients who commission buildings, blurring the distinction between the architect and the rest of the construction team. Contractors have, in some cases become clients. The stakeholders, from clients to designers, contractors, sub-contractors and suppliers, are now multi-national. The built environment is more complex and the environmental performance of buildings, their carbon footprint, their use of scarce resources in manufacture and use is more challenging technically than before. Design decisions cannot be taken in isolation or be based on precedent.

These pressures have led to the growth of a new set of professionals, from project managers who may manage the client relationship between supply and demand that was once the domain of architects, to environmental specialists concerned with the sustainability of construction processes and buildings in use. Mainstream construction is more complex and the trend towards long-term partnerships that deliver a series of buildings has resulted in supplier-led teams, typically led by a contracting organisation rather than the design team.

Where the professions have come from and where they are going is summed up well by the views of Norman Foster given to an audience of engineers:

'I've been very fortunate to work with the best engineering brains of the generation that I am a part of, but in a way we've all had to unlearn something of our past. We've had to readapt in such a way that the specialist engineers understand about their colleagues different and separate engineering disciplines, and can understand and share the vision of a totality. I suppose that, in the language of today, that is called holistic design. As projects get more complicated there is a need for integration, for the dissolving of boundaries.'[22]

CONCLUSION

This more fluid context for architects will lead to opportunities as well as threats. Opportunities are most likely to occur at the boundaries of our silo of knowledge, and by adopting a more flexible approach to how we deploy our professional design and management skills. The architect's holistic role in the design process may or may not be secure. However the architect's skill in balancing the needs of clients with the wider environmental needs of the community are likely to be in greater demand and valued more highly in the future as environmental concerns climb the public's agenda. To return to the model of the professions, architects must continue to do what they do best but will also need to influence government policy, move into management and commerce, and themselves become clients.

FURTHER READING

Although there is a large and growing body of work about the sociology of the professions, very little has been written about architects. The history of architectural practice and the emergence of architecture as a separate profession has been documented by the two studies by Saint and Woods, referred to below. Woods' study explores in detail the emergence of the profession in America, while Saint gives an overview of UK and US experience up to the 1970s. Symes, et al is now a historical piece in its own right, giving a snapshot of the profession in the early 1990s. This period up to the 2010s is an area ripe for future research.

STANDARDS

ARB Architects Code: Standards of Conduct and Practice 2009
www.arb.org.uk

RIBA Code of Conduct 2005
www.architecture.com

The Professional Architect
Brookhouse S *'Part 3 Handbook'* RIBA Publishing 2007

Theories of Professional Behaviour
Macdonald K M *'The Sociology of the Professions'* Sage 1995

History of the Profession
Saint A *'The Image of the Architect'* Yale University Press 1983

Woods M N *'From Craft to Profession'* University of California Press 1999

Architectural Practice
Symes M, Eley J & Seidel A *'Architects and their Practices'* Butterworth Oxford 1995

Cuff D *'Architecture: The Story of Practice'* MIT 1992

Future Practice
Foxell S (ed.) *'The professionals' choice'* CABE 2003

3
IN PRACTICE – 'TOUGH PLAY'[23]

INTRODUCTION

Architectural practice starts with the objective of taking an idea and making it a reality. That idea could be a design to be built, it could be an agenda to be pursued or it could be a different approach to design. But it is not necessarily easy. There are many challenges that all practices face. These include the basic challenge of running a small to medium-sized business enterprise: creating an effective organisation, delivering projects, managing staff resources, sustaining and growing the business, responding to the changing national and international political and economic climate and adapting to structural changes in the construction industry that affect how projects are commissioned. This chapter, with advice and comment from practitioners, explores the primary challenges of practice structure, basic business planning, marketing and managing resources – including staff development. To take your work beyond the studio you will need to manage the processes of design development and procurement which are discussed in later chapters of the book.

ARCHITECTURAL PRACTICE: THE DIFFERENT TYPES OF PROFESSIONAL ORGANISATION

Architects normally have a personal design agenda or a set of principles that drive them. During your formal education you will have developed strong design skills and investigated major design problems and developed innovative design solutions. Architectural practice is the way to apply your skills and to make things happen: it should be the springboard rather than the dampener of your ambitions limited only by your own design skills and resources. Today's major UK architectural practices all started small. Within the professional careers of their founding architects they have grown through a mixture of talent, hard work, good management and good fortune. The catalyst for starting in practice may have been a single design idea developed by one individual but their success has been due to an ability to work as a creative team.

Norman Foster started in architectural practice in 1963 with Su Brumwell, Wendy Cheesman and Richard Rogers. The practice was called 'Team 4'. The practice also included Wendy's sister Georgie who was the only qualified architect but she left after a few months. Like many architectural practices it did not last long. It was dissolved in 1967 but the experience laid the foundations for the careers of two of the UK's most influential practitioners who have gone on to found two major practices: Foster and Partners and Rogers Stirk Harbour + Partners.

The challenge for practice is to build on the skills and agendas of, often strong, individuals and combine these to create a sustainable organisation. This section looks at the four main practice models: the sole practitioner, partnership, the limited company and limited liability partnership. Each model – in different ways – balances the issues of the professional ethos, ownership, management control, liability and succession. Other forms of practice also exist. These include co-operative models where every employee is a member and co-owns the business and trusts, where the ownership is held completely separately for the good of the employees.[24]

THE SOLE PRACTITIONER

A significant number of architects work as sole practitioners. That is not to say that they work alone – but ownership remains with a sole individual. The sole practitioner has complete control over his or her business. She or he may employ other architects, architectural students and support staff but sometimes they will work alone – often from the proverbial 'front room' or 'kitchen table'. One of the advantages is the low costs which come from working from a small office – often shared with other professionals – and the ability to make other life choices such as looking after family or relatives or working part-time that would prevent them from working in a commercial setting. The sole practitioner has total control over the design process although there is a limit to the scale, complexity and number of projects that can be managed effectively.

'I have worked for lots of different types of practice from Farrell's to the BBC. I decided to set up by myself for lots of reasons – some pragmatic, some idealistic and some just selfish – I can be my own boss. After a while in practice I decided that I would not do work that I did not want to or that I did not believe in. It would be unwise to start working by yourself without first developing your business and project skills in a larger practice. We all learn from others. There is no guarantee that you will follow a standard career path.

I like the freedom and the flexibility. I am not obliged to follow one route in my work. I do some research work, I write articles and briefing documents, I enjoy working with 'lay' clients and I actually build too. I also work with other architects and professionals on larger projects so I am not always limited to small scale projects'

Mary Kelly Architect and sole practitioner

The sole practitioner can therefore pursue his or her professional interests – and they have total control over both the projects they take and the decisions that are taken about their professional practice. There will be a trade-off with the scale of projects that can be designed and built – and their ambitions – but for many architects that is a compromise that is worth making.

Sole practitioners may employ other architects but they remain self-employed and the responsibility for the practice remains their own. In effect they are a microcosm of a business and unlike other types of practice are relatively free of statutory control. Because there is no distinction between the individual and the practice as a business they (and possibly their family) are also personally liable for all claims against the practice. Therefore adequate professional indemnity insurance that covers the practice for all possible legal claims is vital if the sole practitioner wants to sleep sound at night.

PARTNERSHIP

Traditionally architects have come together to form partnerships. The partnership model has worked well for many years since the statutory rules governing partnerships were framed in the Partnership Act 1890. The Act defines partnership as: 'the relation that subsists between persons carrying on a business in common with a view to profit.'[25] It has remained reasonably attractive for architects and other professionals for a number of reasons. First, partners still retain their individual status as professionals but work together as partners in the business. Second, it is a flexible business model and relatively easy to create. The partners all enter into a written contract which sets out the detailed terms of the relationship – as well as how the partnership can be dissolved. Third, a partnership retains the concept of the professional as an individual with particular talents and strengths but working together sharing the ownership of the practice. The partnership can be organised along lines agreed amongst the partners who control the decision-making and allocate responsibilities together. This allows for individuality and different agendas to be pursued within the partnership framework. Creative partnerships may therefore accommodate the different strengths – and weaknesses – of the individuals involved.

Because a partnership is a collection of individuals, partners have the liabilities of an individual and also take the added responsibilities of the partnership. Therefore legal action can be taken against a partner jointly (in the name of the partnership) or severally (in the name of the individual partner). Partners are also liable for all the debts of the practice and future obligations in the same way. Whereas in the past this personal as well as joint liability was seen as part of the nature of professionalism it has increasingly been viewed as creating an unreasonable burden on partners without any perceived benefits. Also, as the status of professionals has declined and society at large has become more litigious a partnership has been seen as exposing partners to excessive personal risk.

LIMITED LIABILITY COMPANIES

The majority of commercial organisations operate as limited liability companies. The main reason for this is that its liabilities – unlike the liabilities of partners and sole practitioners – are contained within the company. This happens because the company is recognised as a separate legal 'person', distinct from its owners or employees. The company is liable for its obligations and debts not its owners or directors. You will see that this is a way of limiting the risks of the enterprise and it may come as no surprise that this is the main reason companies have succeeded where partnerships have not as, with some very rare exceptions, partnerships have been unable to take on the risks required as part of its operations. Because the liability of a company is limited – and a failing company may not be able to meet the liabilities it incurs as a business – companies are subject to greater statutory regulation and control. There is substantial legislation covering limited companies, the latest being the Companies Act 2006.

A limited liability company is owned by its shareholders. In turn the company employs directors and other personnel to manage the company and its operations. Ownership and management need not go hand-in-hand. This means that companies can raise money by selling shares in the business to people who are not involved in the day-to-day running of the business. However, in the context of an architectural practice, the directors normally own shares in the company and very few architectural practices raise money by selling shares outside the immediate circle of directors. It does, though, allow the company to employ key personnel for their skills without them having a stake in the company.

It also gives the company the flexibility to offer shares to senior personnel or transfer shares between shareholders. The company itself can also buy shares from directors when they leave or retire. Because of the financial risks associated with limited liability companies where the debts and obligations are held by the company, directors are regulated by the law in a way that partners are not. As with any law if they break the statutory rules they may be prosecuted and fined or imprisoned or banned from being a company director in the future.

Under the RIBA Code of Conduct chartered members who are company directors remain liable for their professional actions and cannot 'hide' behind the company as a separate legal entity for disciplinary purposes. Professional Indemnity insurance is also required for the company and in certain instances directors can be sued in person. The flexibility in ownership, management and control as well as the liability for the company's debts and other obligations resting with the company rather than individual directors make limited liability companies a much more reasonable choice for architectural practices. They also have the advantage of being recognised internationally which is becoming increasingly important. The flexibility of share ownership makes succession more straightforward too.

'We did not want the joint and several liability problems that come with being in a partnership. Being a director limits your personal liability for possible claims against the company. The three of us started the business together and we are equal shareholders. The ownership structure allows you to bring on senior staff and give them the opportunity to take a share in the business. It has not happened yet but that will help our succession planning.

Our image is important and I think being a company is important to our corporate clients.'

Mike Montuschi RIBA Director HFM Architects Limited

LIMITED LIABILITY PARTNERSHIPS

A limited liability partnership (LLP) aims to combine the flexibility of a professional partnership with the protection of a limited liability company. The LLP retains the flexibility of the partnership structure but is a separate legal entity. Instead of partners, an LLP has members. The LLP is responsible for its debts and obligations and unlike a partnership the liability of its members is limited. The LLP is attractive to some architects because it can retain the idea of bringing individual professionals together and also preserve the values of a partnership. It is therefore more attractive to established partnerships that want to limit their liability, and management structure than it is to new architectural practices.

'We changed from a partnership to an LLP because it was more straightforward than forming a limited company. We were concerned about our liability as partners going forward and this was a way of limiting liability without fundamentally changing the structure of the practice. We notified all of our clients about the change, of course, and the limited feedback we did receive was very positive. I do not think clients worry about how the practice is formed – they just want a professional service.'

Nicholas Brill RIBA Brill & Owen LLP

'Changing from a partnership to a LLP was a really easy decision for us. It does not change the culture of the practice but it does address the problem of joint and several liability. It is arcane to go after someone's personal assets.'

Stefanie Fischer Principal BFF Architects LLP

Because of the increased commercial risk LLPs are more highly regulated than partnerships through the Limited Liability Partnership Act 2000. Unlike partnerships, whose financial position may remain private, LLPs have to file financial statements annually with Companies House. These statements are available to the public as part of the increased accountability that goes with limiting liability.

'The privacy issues that surround partnerships become irrelevant because clients – especially in the public sector – will ask for three years' financial statements anyway. Being an LLP is a much more transparent way to deal with a client.'

Stefanie Fischer Principal BFF Architects LLP

OTHER TYPES OF PRACTICE

Architectural practices can also adopt other legal ownership structures. In theory practices can divorce ownership from management entirely – like major companies – by selling shares openly to the public. This form of architectural practice has had a chequered history: compared with many other businesses architectural practice is relatively risky. Those practices that have raised money publically have tended to suffer as a result of the expectations of shareholders who do not manage the company and only want a financial return and the costs associated with public ownership. Commercial shareholders traditionally have a short-term view – in contrast with the lifetime endeavour of professional architects.

An alternative way of creating a sense of belonging for all employees is the creation of a trust that owns the business on behalf of its employees. A trust is also a separate legal entity but unlike a company pays dividends to employees rather than shareholders. The John Lewis Partnership is the best-known example and MAKE, the architectural practice created by Ken Shuttleworth is an example of an architectural practice set up as a limited company but whose shares are held in trust for the benefit of its employees.

STARTING A BUSINESS

Many, many businesses fail within twelve to eighteen month of forming. All started with high hopes and a 'great idea'. Architectural practices do fail but more often than not they manage to survive. The factors favouring success include the high entry barriers to the profession, an in-depth understanding of the market for architectural services, the high value of the service you offer and the risk-averse nature of experienced professionals who are unlikely to be overtaken by a wave of optimism. Also, architects are rarely driven purely by financial gain. We tend to be cautious and realistic about the short- and medium-term prospects for the business. In the same way that your professional advice is valued by clients it is essential that you take professional financial and legal advice. There are risks in running an architectural practice but there are also many rewards – mostly intangible and non-financial. These include having total control over your design work and seeing it built but also growing the business, working with like-minded friends and colleagues, employing and mentoring staff and, possibly, enjoying recognition from your peers and the public for your work.

The key activities for any new business are: 1. Planning, 2. Launching, 3. Running and 4. Growing.[26] (You could also add a fifth: 'exiting' or succession but – as has been said above – architects tend to take a long-term view of both their careers and their business.)

PLANNING

As professional designers one of our key functions is imagining, planning and creating futures for clients and users. Financial planning is an essential part of any business start-up and is a process that continues as an integral business activity. It will include the same activities of scenario-setting and testing of any design process. As with any design project exercise where you investigate spatial or physical constraints and budget, financial planning works with financial constraints of estimated costs, including start-up costs and estimated revenue over a set of time periods to arrive at a rational view of the financial needs of the business. If they cannot be met by either money you will risk putting into the business yourself, money that the banks are prepared to lend you and the fee income you plan to generate then the business will fail before it starts.

The catalyst for starting a practice will usually be the opportunity to design a project or projects combined with a strong desire to do things differently or better or to develop a particular approach to design. Launching a business also means making sure that you choose the most appropriate business model and formalising it. It may require renting office space, buying equipment and software and arranging insurances. It will also mean recruiting, either immediately or in the near future, creative and administrative staff. Lastly, you will need to let the world know you exist utilising all the media available.[27]

RUNNING THE BUSINESS

You will have started a business because you wanted to pursue your vocation: designing buildings. Running a small business, though, also requires new knowledge and a new skillset. As well as designing and managing project delivery – your core business activity – you will need to engage with three other key business functions:

1. Managing our clients: Building and sustaining client relationships.

2. Managing cashflow and resources.

3. Managing people: Leading and motivating your team.

1. Managing our clients: Building and sustaining client relationships

Professional as well as commercial standards apply to the way you run your practice. These act as an overlay to architectural practice. Some of the professional requirements set out by the ARB and RIBA makes good practice sense and others are more onerous. These standards apply to you as an architect rather than your organisation and these personal professional duties apply regardless of the type of practice.

Client loyalty is essential to the success of any architectural practice. We develop long-term and close professional relationships with our clients over the lifetime of a project. (You can contrast this with a chartered accountant who may only meet his clients once or twice a year.) Managing this relationship is an integral part of design and project management discussed in the next chapter. It is important to recognise the potential value your clients hold as well as their current value in sustaining the practice workload. A few well-worn phrases are worth remembering: 'My best award is when a client comes back.' And 'A satisfied client is your best advocate'. Because our business is built on developing a professional relationship it is essential that you communicate with your clients and manage their expectations about project delivery. 'Briefing' is an integral part of this and, again, will be discussed in the next chapter.

Key to this is the appointment contract you have with your client. This is one of the building blocks of your practice and forms the basis for your income too. Your appointment agreement is a contract and should crystallise the client-architect relationship as well as set out the scope of your work and the agreed fee for your professional services as well as a number of other key requirements.[28] Luckily these requirements make sound business sense. Practitioners all recommend that your agreements allow for monthly payments rather than payment being dependent upon the completion of RIBA Work Stages.

2. Managing cash flow and resources

Cash is the lifeblood of any business. Managing cash flow efficiently and effectively is fundamental to the success of any practice. At the heart of this is your contract with your client which sets out how much you are going to be paid and when. You should not under-estimate the challenge of managing cash flow in practice. You will be relying on relatively large sums of money to be paid at specified times and be committed to paying out regular sums of money for rent, loans and salaries, for example, regardless of whether you are being paid by your clients. Therefore you will need a sum of money to cover this shortfall. This is called 'working capital'. It covers the gap between money coming in and money going out. You will see that the faster you can be paid the lower this sum of money needs to be. Some large consumer companies can obtain payment from customers before they pay their suppliers. Their working capital requirement is therefore nil – they may even have funds to spare which they can spend elsewhere. Sadly there is normally a time lag between carrying out our professional work and being paid for it. The aim should be to keep this period to a minimum. You should also realise that it is part of your professionalism to charge the right amount at the agreed time. The obvious way is to charge monthly – even so you will have to wait for payment. It requires time and effort to manage the process, good record-keeping to enable access to accurate financial information fast and then time to monitor your income. The speed at which your clients pay is also important. Chasing them may be a delicate task but it is necessary. Businesses generally fail not because their order books are empty but because their cash flow has dried up. The profession again requires that we manage our resources responsibly. This is in addition to any commercial and statutory requirements.

'Positive cash flow is absolutely critical – especially in the current climate. We aim to set up our fee contracts to generate a monthly cash flow. Otherwise, how can most SMEs survive with limited access to bank credit? Cash flow will make or break the business.

There is a huge risk factor with linking payment to RIBA work stages – especially when the completion of some stages such as planning decisions or tendering are out of your control. You can wait months before you can even send an invoice.'

Stefanie Fischer Principal, BFF Architects LLP

'Managing cash flow is essential. We watch our cash flow very closely and formally review our financial position every 14 days and make sure that we invoice monthly.'

Mike Monutschi RIBA Director HFM Architects Ltd.

Managing resources in the office is also essential. A project-based environment such as architecture is very different from process-based businesses such as banking or manufacturing where the workload is comparatively predictable and in many ways routine. Projects by their nature are finite with a beginning, middle and end. (Design and project management are discussed in the next chapter.) An architectural practice is a series of overlapping or sequential projects. The challenge for practice is to smooth the peaks and troughs of work activity and to match these with the resources available in the practice. Resourcing involves matching skills and knowledge as well as the correct number of people to a project. As you will know this is complicated by the range of skills required at different stages in the project lifecycle from design and presentation skills to project management skills.

Planning, monitoring, managing and reviewing the deployment of personnel is a vital practice management activity. At a financial level it is important that the office generates accurate information on how much time is being spent at different stages of the project. If a project is under-resourced it will run into trouble delivering the design and production information on time. If it is over-resourced it may begin to undermine the viability of the project and ultimately the financial stability of the practice. Also, a practice usually wants to do more than 'balance the books' and to meet its current financial obligations. It will want to reward staff and plan for future growth, renewal of equipment, updating software and set aside money for future investment. It can only do this if it makes a surplus from its core business – designing and managing projects.

In addition to the normal financial discipline that should apply to managing resources the profession – through the Codes and Standards – also makes it a professional requirement to manage resources responsibly and to work in the interest of our clients.[29]

3. Managing people: Leading and motivating your team

Architecture is a knowledge-based business. Your value is directly linked to your competence to practice effectively as well as efficiently. Your knowledge, skills, talents and experience are your greatest assets. Seasoned practitioners joke that their greatest assets go up and down in the lift – in other words – their staff. Even in a small practice you will need to manage a project team and support staff. What does the sole practitioner do?

'Continuous professional development (CPD) is essential. I go to things – seminars and meetings. Because I am interested in low energy building and self-build I join groups that share and promote those interests. It is good for networking too.'

Mary Kelly Architect and sole practitioner

Keeping current and competent is as important for the sole practitioner as it is for any other type of practice. In many ways, because of the solitary nature of sole practice it is more difficult therefore it is essential that the sole practitioner keeps up to date and gets out of the office to share and update knowledge and experience. At a minimum it allows sole practitioners to benchmark their knowledge against others.

As you might expect, professional standards apply. The ARB requires you to be competent to carry out the work you do. (This may – for a sole practitioner – mean limiting the type of work you take on.) The RIBA makes CPD a requirement for all Chartered members and sets out a minimum number of hours per year based on a core CPD curriculum. This makes sound sense to the practice too.

As a relatively junior professional your need for additional support will be greater as you will have both limited experience and knowledge – although you will have many other strengths. If you are working beyond your limits without support or mentoring your motivation as well as your effectiveness will be adversely affected. If you are working in practice before taking the RIBA Part 3 Examination you should be mentored and be given a number of study days to help you prepare for practice. If used correctly the RIBA pedr (See www.Pedr.co.uk) is an effective way to track your professional development and to enter into a regular dialogue with your mentor about your progress, strengths and weaknesses.

'Keeping a good atmosphere in the office is very important. As well as CPD other offline activities and entertainment is important. It helps to build the team and it's fun too. There is only so much a small practice can offer in terms of staff promotion. Bonuses help because they recognise the value of staff to the success of the office in general or a particular project but money is not everything and it is also important to recognise the worth of people as professional individuals.

We believe very strongly that we must make sure our knowledge and skills are kept up-to-date. We organise in-house events and seminars and keep a look-out for conferences and workshops that are relevant. We mentor all new staff and encourage them to take Part 3 and then support them.'

Mike Montuschi RIBA Director HFM Architects Limited

SUSTAINING AND GROWING THE BUSINESS

Growing – or maintaining – the business should be seen as an integral part of how the practice is run and managed. In a project-based environment you need to generate new work in order to stand still. As well as your core business activity – architecture – you must also carry out three basic activities: 1. Planning the business future; 2. Continuous improvement; and 3. Managing change.

You may decide that you do not want to grow the business: the trade-off for taking on larger projects or more smaller to medium-sized projects is greater business risk, more capital, more personnel, potentially a different management structure and culture and a lack of design control. You will need larger office space and unfortunately this normally involves a step-change in your costs. You will need to take on more office space than you need immediately with the aim of expanding into it over the medium-term. In turn you will reach a new plateau of costs resulting in a need to generate a larger and steadier flow of fee income in order to maintain a financial equilibrium. For these reasons many practices make a conscious decision to remain relatively small. The market you are in – possibly geographically if you are a 'local practice' – may also be restricted and this, in turn, will determine the optimum practice size.

PLANNING THE BUSINESS FUTURE

This is the equivalent of the 'start up' phase. It requires setting objectives, reviewing finance for existing and projected needs then reviewing and planning future resources: space, personnel, support services, IT and everything else that aids the key business activity. You will see that this business planning and reviewing your objectives will help you to decide whether your future will be to maintain a stable base, grow – or possibly down-size the business.

Marketing and communication are an integral part of business planning. This should not be confused with advertising which is just one aspect of communication. Marketing involves understanding where you are relative to your competition – other architects (including the ones you studied architecture with) and other members of the design team. Do not forget the clients that had faith in you originally. Marketing will involve maintaining good relationships with your existing clients and developing new client relationships. You should make the most of the available media and use it appropriately. Marketing professional services such as design is a specialist activity. The starting point is to be clear about your strengths and what you can offer that is of value to your clients. (This may be different from what you think your values are – talking to your clients and asking them what they think is an essential part of knowing what to communicate.) You then need to develop a plan that increases awareness of your practice. This is difficult for a small company where funds are limited and you generally do not have the key skills required.

SWOT ANALYSIS

One of the standard business techniques for understanding your position in the market is 'SWOT analysis'. SWOT stands for Strengths, Weaknesses, Opportunities and Threats. The Strengths and Weaknesses identify *'internal'* characteristics under your direct control and the Opportunities and Threats identify *'external'* factors normally beyond your control that will affect the business. For example, your Strengths might include your particular professional design skills such as your knowledge of low carbon technologies or CGI. You may have begun to develop a reputation for your design skills by winning competitions or publishing your work too. One of your Weaknesses may be that you do not effectively communicate your achievements or you have identified, as part of your review of the business, a problem in monitoring and controlling projects.

There will be Opportunities – even in an economic downturn – as well as threats to your practice. For example, BIM may offer an opportunity to innovate. (It may also be a Weakness if you fail to invest in the skills and software.) Threats are more difficult. These may be the state of the economy or the international economic climate. One way of systematically reviewing these is to consider the 'STEEP' factors. These are the Social, Technological, Economic, Environmental and Political factors that influence all businesses either directly or indirectly. Some of these are structural and long-term, some are short-term. An example of a political decision that may affect you directly is regulatory: the government may decide to make changes to the planning system. Or it could be (indirectly) environmental – they may decide to provide grant aid for green building technologies.

CONTINUOUS IMPROVEMENT

One of the principles of marketing is to maintain and develop your distinctiveness. Most practices occupy a middle ground where they compete for the same types of work. If your practice has a specialism or a 'uniqueness' in marketing terms you occupy a 'niche'. The business advantage of being in a 'niche' is that you have less competition. This means that your skills are worth more and your distinctiveness sets you apart from other architectural practices. Being a niche practice can be very beneficial but others will soon try to occupy the same niche – and increase competition. You should look to other professional services for good practice as well as architecture. For example, the success in the advertising industry – a well-established business sector – now depends upon exploiting the opportunities offered by digital media:

'We talk about enduring for decades with the same culture and values and the sense of team collaboration, and you can actually look to Hollywood for inspiration…. You've got to reinvent to remain relevant, and although you have to inject a sense of relentless modernity into everything you do, you need to have a sense of stability as well…'

Ajaz Ahmed co- founder of AKQA, the world's biggest independent digital advertising agency.[30]

There are a number of ways of staying ahead. These include introducing quality systems such as ISO 9001, committing to Investors in People as a structured way of developing and maintaining talent, skills and competence and benchmarking your practice against others using KPIs: key performance indicators.

MANAGING CHANGE: FUTURE PRACTICE

Managing change involves both internal and external review. If you speak to any architect who has been in practice for five or ten years they will say that the practice environment has changed significantly. Change is continuous – it may be gradual or it may be sudden – especially economic and political change which tend to be major. For example, a recent government decided to upgrade every secondary school in the country under the 'Building Schools for the Future' (BSF) programme. Many architectural practices grew to exploit this opportunity. Following the general election the programme was reviewed and most projects were scrapped – overnight. The worldwide economic crisis of 2008 reversed decade-long economic growth throughout Europe and the USA which directly affected the construction industry.

It is difficult to predict sudden structural changes in the world economy but practices must be alert to both forces that will affect the business adversely and opportunities that may also occur. The key success factor is the ability to respond quickly and decisively. This requires a constant process of review and realistic business planning. Sometimes the choices you have to make are not comfortable ones but they may be necessary for business survival as well as growth.

CONCLUSIONS

Architecture gives you the opportunity to share in running a business or to run your own business. To do so will require combining your professional design skills with new-found business skills. The majority of architects will be involved in running an architectural practice at various stages in their careers.

This chapter has introduced a number of key themes. In deciding what type of business structure to choose a central concern for practitioners is business risk and personal liability. Many think that the idea of 'partnership' where individual professionals come together to practice architecture is outdated – the personal risks outweigh the flexibility. The ethos and culture of partnership is retained in an LLP but personal liability is limited by its legal structure. The majority of businesses operate as limited liability companies – and practitioners see benefits as well as limiting personal liability. These include international recognition.

Running a business requires a different set of functions that are separate from your key design skills and the core function of practice: making architecture. These functions must become an integral part of your practice. They are evaluative and iterative. They require planning and reviewing. The business that excludes these activities – however talented its professional personnel – will either fail outright or fail to achieve its potential. In a knowledge-based profession retaining and supporting professional and support staff is also key to your success. It is therefore important to value staff contribution to personal career planning and to refresh the skill and knowledge base of your practice. Sometimes you will also have to 'buy in' additional professional skills such as finance and marketing at appropriate times.

In addition there is a professional, regulatory overlay to your practice – both as an individual architect and as an architectural practice. Both the ARB Code and Standards and the RIBA Code of Practice apply not only to your professional work but also to the way you conduct your business. Therefore, although your personal financial liability may be limited by the form of practice (LLP or Limited Liability Company), you will not be able to avoid your professional responsibilities set out in these Codes.

Continuous change is also another theme. Managing change and continuous improvement are integral business activities. Practitioners interviewed all refer to change. Although you will have achieved a level of professional competence through your education at university and professional development in practice you will need to continuously develop and refresh your knowledge throughout your career in order to adapt to the changes in the industries and economies we work in and the architectural profession.

FURTHER READING.

Chapell D *'The Architect in Practice'* (10th ed.) Wiley-Blackwell 2010

Foxell S *'Starting a Practice'* RIBA Publishing 2006

Ostime N *'Architects Handbook of Practice Management'* (8th ed.) RIBA Publishing 2010

4

MANAGING DESIGN – ENGAGING THE TEAM

'Design Management is the discipline of planning, organizing and managing the design process to bring about the successful completion of specific goals and objectives.'[31]

INTRODUCTION

Design management is one of the core activities of any architectural practice. (If you take away the design projects then the practice is nothing. If you manage them badly you will go out of business.) It follows that managing design projects effectively – from inception to completion – is the key to a practice's success and its viability as a business. To take your design work 'beyond the studio', and get it built, you need to understand and engage with multiple stakeholders – from clients to contractors – sometimes in different ways depending on their attitude to risk and their experience of the design and construction process. The chapter on procurement shows that our core activities sit within a highly complex and sometimes fluid environment where stakeholders – including the architect – may adopt different roles. Effective management of the design process is an integral part of the procurement process but is, in itself, a complex set of activities. This complex environment sets out a number of challenges for the architect.

'Design management wraps around the design process. All the inputs and the outputs have to be managed for the design to be successfully delivered. How can you manage an iterative process [like design] and manage a complex design team?'

Dale Sinclair Dyer Architects

The aim of this chapter is to help you to navigate through these complexities and the range of challenges for the architect by introducing the key principles of management and some of the tools and techniques that apply to successful design management. The topics include:

■   an introduction to management as a separate discipline and profession,

■   an overview of project management methodologies generally and the RIBA Outline Plan of Work, (one of the standard project management models used by the construction industry),

■   the challenge of working with teams,

■   the use of BIM as an integrated software model for managing the information generated by project stakeholders, and

■   the use of briefing as a design management method that runs through the lifecycle of the project.

The guide to further reading at the end of the chapter will allow you to develop your knowledge and interest in this critical area of professional architectural practice.

THE GROWTH OF MANAGEMENT AS AN ACADEMIC DISCIPLINE AND PROFESSION

Architecture has its own 'silo of knowledge', a tradition of critical theory and an historical and technical context often demonstrated by the end *product* – the building. However, architectural design as a creative *process* which, by its nature, is also highly iterative. Design management is the system used to move architecture from a process to a product using finite resources. If design management is essential to successful architectural practice then it follows that you need to understand the way the design process can be effectively managed. As one architect-design manager put it:

'You have to design the process as well as the product'[32]

Alan Brown Initiatives in Design Architects

Management theory grew out of two separate disciplines: the 'harder' discipline of engineering (mechanical, civil and structural) including industrial statistics and the 'softer' discipline of the social sciences as applied to organisations – how individuals perform physically (our productivity) and behave psychologically (our motivation), and finally how we operate in teams and construct business enterprises (organisational theory).

Both theoretical strands are founded on the scientific process where models or theories are constructed and tested based on multiple observations. It follows that if an accurate model can be constructed then this can be adapted and possibly tested to propose new futures. This requires a structured, analytical approach to problem-solving. Typically events or problems are broken down into elements and then reconstructed as a theory with which to model future broadly similar 'what-if' scenarios. The ability to accurately predict outcomes based on past performance is a powerful management technique: in an uncertain business and project environment it allows you to forecast and evaluate multiple outcomes and to manage risk.

Modelling human interaction in groups and formal or informal organisations is more difficult and remains a major preoccupation of organisational sociologists and psychologists. Routine, repetitive, sequential tasks can be observed relatively easily but the more fluid knowledge-based activities and the transient relationships that characterise design project teams are more difficult to observe and model – hence an enduring curiosity for organisational theorists in the creative design process. Management professionals seek to use and add to their own 'silos of knowledge' and apply theory in the workplace with the aim of achieving competitive advantage. As a separate profession management consultancy has grown, in a relatively short period, to be a major force of influence across all business areas.

From the vast and growing body of management theory two areas are particularly relevant to design projects: 1. the 'hard' system-based approach to modelling project processes; and 2. the 'soft' approach to managing teams and motivating individual professionals that happen to work in these project teams.

For a process to be well-managed it must have three phases: 1. *'Plan'*; 2. *'Action'*; and 3. *'Review'*. Observing the professional project environment of architectural practice you will normally reach the conclusion that architects spend most of their time 'doing' (the *'Action'* phase). We will do some *planning* (the first phase) – but probably not enough – and very rarely carry out a review (the final phase) of a project. If we do it tends to concentrate on the 'product' rather than the 'process'. Often the *review* phase is carried out in the context of failure – either technical, budgetary or performance resulting in programme delay – or a combination of all three.

FIGURE 4.1

**Circular diagram of
the basic management
process**

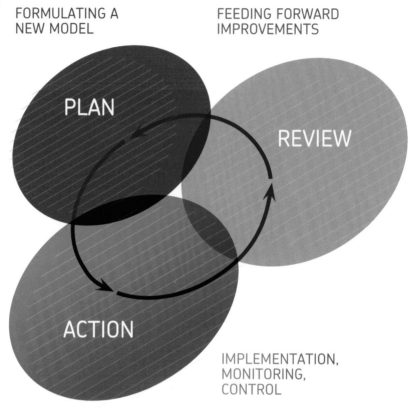

FORMULATING A
NEW MODEL

FEEDING FORWARD
IMPROVEMENTS

PLAN

REVIEW

ACTION

IMPLEMENTATION,
MONITORING,
CONTROL

Management theorists and practitioners will normally start with the 'review' phase
– collecting information in a structured, detailed and systematic way in order
to construct theories – 'to plan' – and then to put these into action – or 'test'.
To complete the cycle requires monitoring and gathering further information in
order to exercise control. This process is integral to effective design management
generally and also 'briefing' which will be discussed later in the chapter. Systems
analysis – where processes or events are broken down into elements to see how
they interact – is a key management method. The objective is systems synthesis:
to design a better system by re-assembling, modifying, amending or deleting
components. We accept this as a valuable management method.

Two weeks before the London 2012 Olympics Games started the company responsible for security, G4S, admitted that they were unable to provide all the security guards required. They had originally been asked to provide 2,400 guards but the Games organisers increased this to 10,400 and as the Games approached they were 3,500 under strength. The armed forces were called in to make up the shortfall.

Parliament branded the process as a 'shambles': 'The largest security company in the world… turned years of carefully laid preparations into an eleventh hour fiasco.' In response, G4S commissioned a management consultancy, PwC, to find out what went wrong. They used a systems approach: breaking the problem down into its parts – starting with the request by LOCOG to more than quadruple the number of guards required for the Games. PwC concluded that:

'The Olympic contract was unique in terms of scale and complexity, but the Company was capable of fulfilling the contract; the issue was in its delivery…The monitoring and tracking of the security workforce, management information and the project management framework and practices were ineffective to address the scale, complexities and dependencies of the Olympic contract.'[33]

Project management, briefing, information management, monitoring and tracking are all key components of effective design management.

APPLYING METHODS TO PROJECT MANAGEMENT

To understand how to manage a type of project and know how to control the processes involved, information from similar projects must be collected, modelled and monitored. The more comprehensive the information or data-gathering phase, the more accurate the model and the greater the likelihood of effective project control. Because design and construction processes are particularly fluid – and projects are costly – any methodology that can reduce uncertainty and therefore the risk of failure in terms of the process and performance has to be of value to stakeholders, especially if they are risk averse, which most stakeholders are.

There are many proprietary software project management tools available but all are based around the graphic representation of a project as a linear process in the form of a 'bar' or 'Gantt chart'. The Gantt chart is, surprisingly, a relatively new project management tool – gaining prominence in the Polaris nuclear submarine programme in the 1950s.

You do not need to understand the underlying mathematical theory – network analysis – that underpins the bar chart but it is important to understand that it is simply a graphic representation of potentially endless often inter-dependent events. The particular value of the Gantt chart in design and project management is that it can model different activities over time and identify the dependency of one activity on another as a linear process. The shortest linear dependency from beginning to end is known as the 'critical path'. Activities off the critical path have an element of time flexibility – or 'float'. The value of the Gantt chart is that you can monitor progress and the effect of the delay – or acceleration – of one activity on the whole project. You can also model activities and create different scenarios relatively quickly. As well as the two variables shown on the Gantt chart – activity against time – you can also monitor and model the resources required to carry out particular tasks or activities. By assigning values to these variables (time and resource) you can predict the resources required to complete project tasks. By continuously monitoring these variables you can measure actual performance and quickly compare this with the predicted performance – thus changing estimated completion dates or the resources required. We carry out these activities intuitively – almost subconsciously – in our own design work. But this informal approach cannot survive in the wider project environment where we are interacting with other professionals.

Construction projects depend on Gantt charts for monitoring and controlling performance. The chart shows the inter-dependence of events in a linear format and this accurately reflects most construction processes – digging foundations, pouring concrete, laying bricks, installing ceilings and so on. Each activity follows on from another. Sometimes they can be contiguous and in series or they can run in parallel.

Although complex, a linear format is relatively straightforward. Design, however, is an iterative process. How can it be programmed?

THE RIBA OUTLINE PLAN OF WORK.

Perhaps because design is an iterative, fluid activity that is difficult to describe to other stakeholders – or because the architect was traditionally the design project team leader – the architectural profession developed a standardised model for design and construction projects – the RIBA Outline Plan of Work[34] (OPoW) was first published fifty years ago in 1963. The construction industry was very different then: general contracting was the predominant procurement route, for projects other than basic industrial buildings. The architect was the main – possibly sole-point of contact with key stakeholders: managing the client relationship, leading the professional design team and managing the construction contract and key sub-contractors. Projects followed the linear, 'baton-passing', sequential path that characterises general contracting. The OPoW was – in its original and enduring form – a reasonably accurate model of the design and project management process.[35]

In its normal form the OPoW defines different, sequential, project phases from early investigations, strategic briefing, sketch design to detailed design, regulatory requirements such as planning permission and building regulations approval following design freeze, tendering and contractor selection, the construction phase and project close. It also sets out what different stakeholders should be doing at different phases of the project and key milestones or points where key decisions (such as design freeze) should be taken.

Over the last fifty years the project environment has changed significantly: projects have become more complex, attitude to risk have changed and procurement routes have multiplied. The roles in the design team have also multiplied with specialist project managers sometimes leading the project and the client represented by specialist design managers. As well as changes in the architect-client relationship it is no longer a 'given' that the architect will lead the design team or be involved in the construction phase of a project. Client expectations have also changed, in short there is an expectation that all phases of the project will be managed more effectively with the aim of reducing risk and creating more value from the available resources.

As a result of continuous project scrutiny and review where clients on the 'demand-side' require greater value and reduced risk and increased domestic and international competition 'supply-side' contractor stakeholders developed more innovative procurement routes. This increased complexity also promoted project management as a separate construction management discipline. It became increasingly difficult to create a single design and project management model that applied to all projects. In response to changes in procurement the OPoW has evolved to take account of the different roles the architect takes – and in particular that of Lead Designer – and the different ways that design processes can be managed.

However, the traditional OPoW sequential model remains relevant and appropriate to projects that follow the general contracting procurement route. All projects also share certain key elements – especially in the early stages. The OPoW has also been adopted by other members of the design team and as a result the team can immediately relate to key decision-making points in the project management cycle, for example where the detailed design is approved and where further changes will have significant impact on the design, budget and timescales. A further criticism of the OPoW is a structural one: it defines a particular project model, freezes the architect's roles in time and possibly excludes other activities that the architect can perform.

The OPoW has certainly been an effective tool for project planning and for monitoring and controlling the *'action'* phases of effective project management. It should also be used as a vehicle for the *'review'* phase – the key to improving the performance of the architect during the key design, production and management phases of a design project.

Different approaches to project management roles and responsibilities that better reflect the various complex procurement methods in use today coupled with the ease with which bespoke project management software tools can generate new models has led to a proliferation of project management models. Ultimately they all rely upon the same mathematical processes and algorithms originally developed in the 1950s. They satisfy a need to reliably model data and to monitor the varied activities in design and construction with the aim of more effective project control.

It should always be remembered that the original inputs as well as their evaluation are subjective even though they take on an objective, measurable importance. Ultimately they rely upon the experience and judgment of the project team to effectively manage the process. This judgment can be improved by learning from experience and using project management models as a way of historically evaluating ('reviewing') different projects in a standardised way.

Project management methods have introduced a more formal approach to how projects that can be modelled, measured and ultimately controlled. This is attractive because it helps us to manage project uncertainty and reduce risk. But project management methods are only a tool: the way teams come together and interact is arguably the key to project success.

BUILDING INFORMATION MODELLING (BIM)

'On average, we re-draw buildings 3.4 times.'[36] This was one of the results of a study into the design and construction process and the collaboration required between architects, consultants, sub-contractors and contractors. The revision rate is due to the changing inputs of members of the design and construction team as they finalise production information. Although design is an iterative process the inefficiencies created by the uncoordinated inputs of different members of the team can result in design ideas being dissipated and errors that are only discovered at the construction stage. Building Information Modelling (BIM) aims to improve efficiency through greater transparent collaboration by all members of the design and construction team. It does this by creating a virtual 3-D model of a building that can, depending on the procurement route, be handed to a contractor or that the contractor can contribute to and, when the building is completed, can be passed to the owner and user. Conceptually, it spans the whole lifecycle of a building but its greatest value in managing the design process is the level of collaboration and teamwork it allows for the design team.

'BIM is a process. It is underpinned by technology but is also about a collaborative approach. 3D modelling is not only about the physical, tangible evidence created from that process, rather an iterative process that begins to cover operation, maintenance and eventual decommissioning of a building and far exceeding what the building may have been if designed only in 2D. What we understand and value is that the design process and the idea behind it, albeit logical or theoretical, are still the same. The tools and their BIM environment do not change how we design buildings just the integrity of the information we use to document the construction of them'.

Rebecca de Cicco, KSS Architects[37]

BIM is important also because the government, having seen the potential efficiencies that BIM can create, has made it a cornerstone of the Government Construction Strategy with the aim that all government departments, as key stakeholders, will adopt BIM to improve collaboration with the design team and contractors. Through BIM the government also seeks to achieve the Strategy's key objective of reducing the cost of construction, the carbon burden of constructing and the operating of buildings by twenty per cent.[38]

BIM will be most effective in the design management process if it successfully integrates and manages the design inputs of all members of the design (and construction) team. It will need to operate not as a 'bolt-on' piece of technology but a platform for communication and the collaborative modelling of all information. In doing so, it will also need to mirror the decision-making structure and lines of communication of the design team. Architects have a key role to play as both Lead Designers and design team managers. In order to reap the benefits of BIM architects will need to establish leadership and control of the design process. There are also obstacles to the full integration of design information. These include legal issues such as the liability for design errors in a fully integrated design model. More importantly, inefficiencies will creep in if the authority to manage the design of the BIM model is separated from the decision-making authority of key stakeholders. For example, if design decisions taken by the design team are changed by the client after a key stage in the design development then part of the value of BIM will be lost. The need to effectively manage the design team will still exist.

MANAGING THE PROJECT TEAM

'It is essential that knowledge about a project does not simply reside in the head of the lead consultant or lead designer, but that it is 'downloaded', shared and made available to everyone in the team. If this knowledge can be harnessed and made available to all, it can serve as a design tool, stimulate effective actions and responses among the team, facilitate planning and be used as a means of conveying the status of the design to the client.'[39]

Dale Sinclair

Tools and techniques based on the mathematical analysis of activities and resources over time such as Gantt charts, proprietary project management software packages and BIM are only a part of successful design management. In a project design environment the key resource to be managed is the expertise of the different professional design disciplines required to make the project happen. Managing creative teams and co-ordinating the input and interaction of talented and experienced professionals is complex and difficult in the fluid and fuzzy early stages of a project as well as the detailed design, production information and construction phases. Architects contribute significantly as skilled designers and also as managers of the design input of other professionals.

The structure of design and project team can be explained using stakeholder analysis. Stakeholder roles will vary depending on the procurement route. A typical traditional contracting structure is as follows:

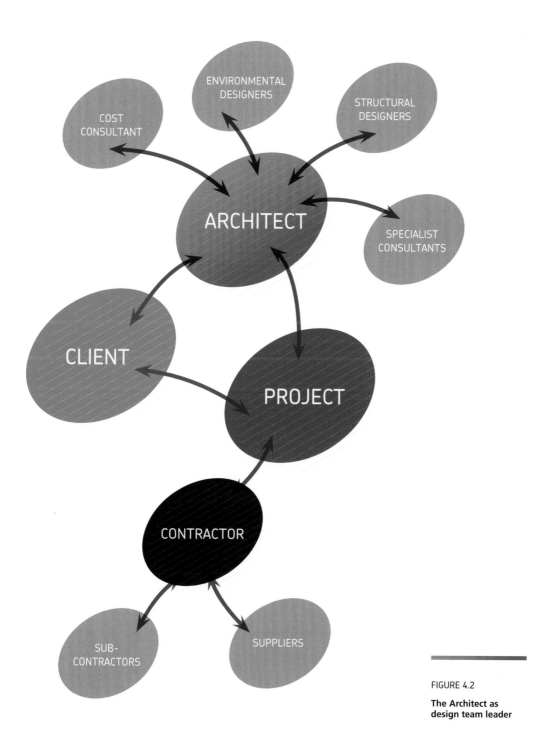

FIGURE 4.2

The Architect as design team leader

Other structures will be designed to suit the role of different stakeholders and as a consequence the role and involvement of the architect will vary too.

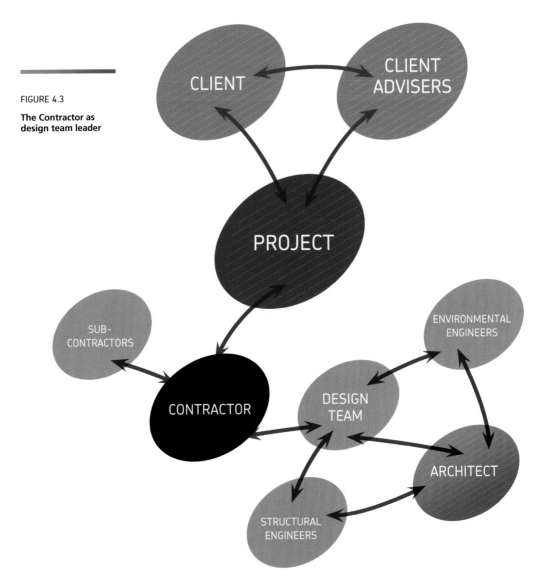

FIGURE 4.3

The Contractor as design team leader

Another way of showing the different roles and responsibilities is to create a design responsibility matrix. Design responsibilities may become blurred, especially when the architect is employing some members of the team as sub-consultants or some design is being carried out by specialist contractors. The challenge for architectural practice is to recognise and adapt to the different roles required by the selected procurement route. Different phases of the project call for different knowledge and skills as well as different numbers of routes and team members. Clients normally expect to see senior architectural personnel leading the project. The design project manager needs to manage *across* different members of the team, *upwards* to the client stakeholders and *beyond* the team to other stakeholders such as specialist suppliers and (sub)-contractors.

The project management skills are:

- personnel and resource planning,

- delegation,

- leadership within and beyond the design team,

- negotiation, and

- communication.

The more complex the project, the more challenging the design and stakeholder management burden. For the architect and design team leader the key task is to produce the right information, at the right time, in the right sequence and to co-ordinate information across the design team to control both design outputs and ensure that information disseminated or published to other stakeholders is correct and error-free. It is worth noting that contractors deploy the same skills in relation to the planning and co-ordination of sub-contractor packages during the construction phase. There is the added problem that the risk of errors has a far greater consequence in terms of cost and project performance. For these reasons the management contractor or design and build contractor may also take a key role in the management of the design as well as the construction depending on the procurement route chosen for the project.

Formal project management methodologies have a key role to play in managing resources and design outputs. On small projects one person – the sole practitioner, for example – may possess all the necessary skills. The project may be sufficiently small enough to allow one person to carry out all tasks – but that is unusual and highly restricting in terms of the size and complexity of project that can be successfully executed. With larger projects, as well as using more people, Gantt charts can be used to plan, monitor and control the deployment of resources.

Reviewing project performance is an essential activity: it allows lessons to be learnt and allows the design project manager to accurately predict the resource requirements in terms of the deployment of appropriate skills as well as numbers of staff and hence predict the cost of resourcing future projects. Architectural practices can build a database of past projects which together with a set of uniform performance indicators ranging from simple measures such as profitability to more complex measures as stakeholder satisfaction can be used as an objective evaluation of past projects and inform the effective management of future projects. Lastly it should be recognised that the management of creative resources is not easy: when things go wrong 'throwing more people' at the problem generally creates more problems and affects the quality of outputs. The error rate goes up and work has to be re-done as design team members drafted in at short notice will either lack knowledge of the project and may also not be as immediately committed to the team. The complexity and skills to manage the iterative rather than linear nature of design cannot be underestimated.

BRIEFING FOR EFFECTIVE DESIGN MANAGEMENT

'Briefing is an evolutionary process of understanding an organisation's needs and resources, and matching these to its objectives and its mission. The difference between this and a brief is that the brief is a product of the process; it formalises the decisions and actions to be taken and may be produced at key stages of the process.'[40]

Briefing is one of the key techniques for effective design management. Broken down into its parts and re-constructed as a visible, explicit process, it can be modelled and used systematically in design management. The RIBA Criteria consistently emphasises the importance of briefing as an essential and necessary skill because it is at the heart of an architect's core activity – design. By extension, any technique that can make the design process more effective will also benefit the practice of architecture.

Briefing has 'hard' and 'soft' aspects. It can be a formal process but it may also involve informal incidents that add significant value to all stages of design – not just inception. 'Hard' terms include objectives and features that can be measured and reviewed at project close: cost, time to completion, physical metrics such as size and performance. It may also ask factual questions such as 'Did the project meet the stated needs of the organisation?' At this level, therefore, the brief can set agreed key indicators of project success: the inputs and the outputs.

Design managers also say that the major benefits of briefing concern the 'soft' aspects of design management. The briefing process provides the forum for communicating and engaging with stakeholders, managing the stakeholder relationship, focusing design effort and managing the evolving design process so as to achieve the 'hard', measurable outputs. It is important to understand the people who direct resources into a project and to have an understanding of their goals and motives as this will help to influence and shape the outcomes of the project.

'Briefing is a process of managing stakeholder engagement. In one sense it is about a conversation with all stakeholders not just the client or the users – and it runs continuously through the project. It is wrong to think that you just hand over the brief to the architect and he or she goes away and gets on with it and comes back with the finished design.'

Alastair Blyth architect and briefing consultant

Experienced architects know that clients – the major stakeholders – tend to react to the designs they are presented with. They also change their minds. Experienced design managers know that 'changing your mind' is part of the creative process and a good – rather than bad – sign of stakeholder engagement. It reflects the iterative nature of the design process itself and stakeholders become actively engaged with the design development.

In order to control and model this iterative and potentially endless process design managers and briefing consultants identify key stages in the project and key points for the brief to be frozen: sometimes referred to as 'The last responsible moment.'[41]

To place briefing within a project structure it is useful to consider a design project in a wider context. Latham, in 'Constructing the Team'[42] divided projects into three broad phases: Pre-project, Project and Post-Project. These stages deliberately set the widest boundaries in the project process. For example, the 'Pre-project' stage will involve multiple project scenarios – possibly without a building design. The project stage takes in the activities described in the RIBA OPoW from design inception to completion and the Post-project phase recognises that buildings are designed to be used and that valuable lessons may be learnt from post-occupancy studies of building performance and evaluation of the entire procurement, design and construction project process. The three phases mirror the 'Plan – Action – Review' phases of the generic management process.

At the **Pre-project** stage the brief will be strategic and built on the client organisation's needs. It should ensure that the client's design objectives align with its corporate objectives. It may involve testing the feasibility of a project in its widest sense using multiple scenarios – one of which might be to not build at all. This testing helps to define priorities and to make aspects of an organisation that may be implicit, explicit: its vision for the future, for example. The Strategic Brief therefore becomes a way of communicating objectives as well as essential project requirements and provides a secure foundation for design development. Design communication at this stage might include bubble diagrams and sketch schemes.

A **Strategic Brief** may include details under the following headings:

- Vision Statement: *to capture the essence of the project.*

- Objectives: *what the project has to achieve: business, organisational, design.*

- Needs: *things required that must satisfy different stakeholders.*

- Expectations: *how these needs are to be met in measurable terms. For example, environmental performance.*

- Growth and Change: *sets out how the building may need to adapt to planned, foreseen change.*

- Priorities and measures for success: *how the project will be judged. For example, key performance indicators: time, budget and value-for-money criteria.*

- Decision Framework: *key milestones.*[43]

At the **Project** stage the brief will be more detailed and define the functional requirements: size, cost and possibly a statement about the image that will make the organisation's vision explicit. The 'Functional Brief' at this stage may include the scheme design and would be approved by users, the client and any advisory team. It may also include statements defining attitudes to the project and construction risk to satisfy funding requirements of different stakeholders such as banks or in public sector funders. Depending on the size and complexity, separate detailed briefs will be prepared. For example, these may be for equipment, environmental conditions or communications. The Functional Brief may have the same headings used in the Strategic Brief but the information will be much more detailed filling out the areas of the Strategic Brief with firm data collected from users and the client. It sets out, in building terms, what the client organisation wants to achieve. Outputs will include design drawings and sketch details that illustrate specific requirements as well as room data sheets. Again, it is the iterative development of design and briefing process.

At the **Post-project** stage the briefing completes the circle and reviews the design and construction process. For example, it may examine how the building fabric performed in practice when measured against the predicted performance. It could survey and record feedback from users. These exercises in establishing the performance of the *product* are perceived as expensive and time-consuming but the costs are minimal when compared with the cost of construction. Valuable lessons can be learnt to feed into future designs. Difficult questions such as 'Did the design work?' may yield uncomfortable answers but we can only improve by asking these questions. Stakeholders should also review the *process* to feed forward into future projects. 'Did we ask the right questions?' 'Were the objectives too ambitious?' 'Were we *too* risk averse?' 'Did we communicate effectively?' This review stage is particularly important because of the finite nature of projects. Stakeholder teams come together for a project and then disperse when they are completed. The same teams may never work together again. Even designers move on and project completion is often the time to make a career move. Without this review of key indicators of project success, rigorous analysis and recording of findings the project lessons may be lost. The process should be seen as positive rather than negative: providing valuable data for future projects. Architects, in particular should evaluate their designs and build a body of knowledge and expertise based on project experience.

This iterative process of stakeholder engagement is very difficult to simulate in the design studio. Schools of architecture will often start a project with a programme or brief. This may evolve as a result of further investigation – including conversations with stakeholders – and further study and research. This process of investigation which informs the design is one of the valuable, key skills architects develop but it is important to emphasise the iterative rather than linear nature of briefing – that it may be the mid-point and the end-point as well as the starting point of the design process.

CONCLUSIONS

This chapter aimed to introduce you to management as a separate discipline, to explain the role of the architect as a major stakeholder in design management and to explore the briefing process. Management as a profession has its own huge and growing 'silos of knowledge' and the management process of 'plan – act – review' has much in common with the design process. It would be easy to say the two are similar and that the creative design process is a management process. However it may be more realistic to look at the differences: the different 'hard' and 'soft' disciplines of the management profession have little in common with the design professions. Architects learn creative design skills in the studio and this learning process is very different from the processes of management. If you recognise this then you can understand that successful design management is something we must carry out in addition to our creative design. Successful design management requires a different set of skills including planning the design management process, managing stakeholder engagement, managing the design team and the project team. Briefing is a valuable tool in design management – but only one of the tools that is required.

Architects work on many different types of project for very different clients. It is natural when discussing design management and briefing to talk about multiple stakeholders: clients, funders and users – large organisations commissioning complex projects that require rigorous design management. It should be remembered that the same process applies to small projects – which are often complex too. Small domestic projects for owner-occupier clients are rarely simple. The client will be user and funder combined with a different approach to project risk. Effective design management and iterative briefing in particular will still be essential for project success.

Lastly, it is worth remembering that key stakeholders, clients, funders – and possibly project managers – will normally be highly-skilled management professionals who will approach projects with a management perspective. They will be intrigued and excited by the design process and the creative design environment but, depending on their role, will deploy their analytical management skills to review not only the design but also the design management process as well.

FURTHER READING

Sinclair D 'Leading the Team: an Architect's Guide to Design Management' RIBA Publishing 2011

Race S 'BIM Demystified' RIBA Publishing 2012

Blyth A & Worthington J 'Managing the Brief for better design' (2nd Edition) Routledge 2010

5

THE ENGLISH LEGAL SYSTEM

INTRODUCTION

The process of setting a building in its environment involves a subtle combination of factors. As an undergraduate student of design you will have responded to a brief to propose a physical solution to a brief or an agenda. However, buildings are also framed by a set of constraints that go beyond the brief. Some are physical and others administrative constraints which are discussed in later chapters. These constraints can only be fully understood in their legal context. In this chapter you will begin to understand the origins of the rules and regulations that control not only the processes that govern design and construction projects but also the way professionals have to engage with their clients and society generally.

The following topics are covered in this chapter:

- The nature of Law.

- Civil and Criminal Systems.

- Common Law, Equity and the role of the State in making new laws.

- International influences – in particular European Law.

THE NATURE OF LAW

Before you look at the English legal system close your eyes and imagine you are on a desert island with a large group of shipwrecked citizens. Being relatively civilised the citizens have already formed an Island Council. You have been tasked by the Council to design not a building but a legal system for the community. Think about the following questions for a few minutes:

■   What do they mean by 'legal system?'

■   What should its objectives be?

■   Where would you start to define what has to be put in place?

■   How would you make sure that it was fair?

■   How can you make it effective?

In coming up with some answers you should begin to understand the problems faced by any society in construction of a fair system of laws.

'WHAT DO THEY MEAN BY A LEGAL SYSTEM?'

It is a fundamental question – but how do you answer it?

Essentially it is a system of rules that the community will respect and follow. It has to have a number of components:

■   guidance and rules on acceptable behaviour,

■   a system for monitoring this guidance or rules,

■   a mechanism for deciding of the rules have been broken that is 'fair', transparent and consistent so that citizens can see when the rules have been broken,

■   a method for deciding what to do to put things right and possibly a punishment that is appropriate and fair – and again consistent,

■   a way of publicising these decisions so that fellow citizens on the island know what has happened,

■   a method of reviewing decisions in case they are wrong,

■   lastly the rules have to be enforceable and to be seen to be being enforced consistently.

Hence the standard definition of the law as:

'The enforceable body of rules that govern any society'[44]

WHAT SHOULD THE OBJECTIVES BE?

A few have been touched on already. These may be some of your possible objectives for your new system.

- Fairness.

- Consistency.

- The substance of the system – the laws – should reflect the moral code of the society.

- Effectiveness – you have to be able to apply them in practice.

- Independence.

- The ability to change if necessary to suit circumstances.

- Enforceability.

You will see that your new system for your desert island quickly breaks down into a number of elements: a set of rules; a method for applying them; and an organisation to enforce them. The challenges are: 1. to make the rules appropriate to your society; 2. to design a decision-making system that is fair and transparent (and possibly independent); 3. to apply those decisions effectively; and 4. to review and change decisions when they are wrong.

WHAT HAS TO BE PUT IN PLACE?

This question is deliberately vague but the same process of design applies here as with a design brief. Where do you start? Having analysed the problem it is always useful to see what other, more experienced designers have done in the past – precedence.

Start with the substance of your rules – 'What will they be?' This is a huge question and one whose scope appears to be constantly expanding. We have an idea of what is right or wrong, a moral code. The origins of that moral code start with family and religion – and this is the starting point for most established legal systems. Although we live in a relatively pluralistic society the English Legal System – and its many forms – follows many of the rules set by the Judaic-Christian tradition. You want a moral code that can be understood and enforced and the Old and New Testaments provide a good starting point. They are reasonably well understood and respected. Both Judaism and Christianity also provided sophisticated systems for administration and enforcement. At its best it is also self-regulating. But not everyone from the shipwreck follows these traditions. There are other religions and also some with no belief at all.

Having decided to borrow a religious code of behaviour for the substance of your new legal system, and possibly its administration (if you had a convenient religious organisation to hand on your desert island), you will quickly realise that this is not enough.

First, there are other problems on this island that need to be resolved which are outside the scope of a moral code of behaviour. For example, disputes between individuals about basic territory and ownership, bargains that are not kept or simple problems that may escalate to the point where they affect your own life – your neighbour is making a noise when you want to sleep at night. If this persists then it will take on a new dimension – possibly as an invasion of your privacy or your right to enjoy a peaceful night without interruption. This and other problems do not involve theft or violence – but they make up the bulk of disputes in society. They are essentially private but have an impact on relationships between individuals and groups. Therefore they may need to be treated differently from theft or violence.

Second, you will realise that this new system is an opportunity for social control. You can devise *new rules* as well as relying on established moral codes. Anyway, you know that these codes are not appropriate for every problem, especially where there is not a moral judgment to make. This new system presents you with a dilemma. 'Do you design a system that is completely independent in substance and administration or one that is controlled by the Island Council?' Or, is there something between the two? What you decide may have an impact on the next two questions.

HOW DO YOU DECIDE THAT THE SYSTEM IS FAIR?

A decision is fair if the same decision applies to similar problems in similar circumstances. Your citizens will therefore be looking for consistency. They will more readily accept the decision as fair if they can understand it – ideally in their own terms and using easy-to-understand everyday language. Also the decision must be commensurate with the scale of the problem. The decision is often a punishment – a long tradition being established in religious texts – but can also be forgiveness. So, if your neighbour steals two of your sheep he should be dealt with more harshly than if he steals only one. In other words the punishment should fit the crime.

As well as crimes such as theft and violence there are also private disputes between individuals which cannot be solved by negotiation or agreement and which need the intervention of a neutral third party to whom you give the authority to make a decision. In return you agree to be bound by that decision. The solution is likely to involve returning property so that you are put in the position you were in before the cause of the dispute occurred – or a financial settlement which makes up for the loss. The advantage of the latter is that it is simple and you can compare similar cases against the same measurement. The resolution of these private disputes is also important for the community: on your island the breakdown of relationships can affect others and may also escalate to crime if not resolved.

However this leaves a range of disputes where money is not an appropriate solution – your noisy neighbour, for example. You hoped to use your system to stop it – not to achieve a financial settlement. Furthermore, no amount of money is going to give you the good night's sleep that you now crave. This is more difficult to resolve.

There is also a range of problems where innocent people suffer through no fault of their own but possibly due to someone else's carelessness – 'negligence'– getting run over by mistake or falling off a balcony because the handrail was not secure.

In the close-knit island community spreading malicious rumours, 'slander', will have a significant effect on the way your island society interacts as well as causing distress to the individuals concerned.

Then there are the laws that the Island Council cannot resist making as an instrument of social control. These laws may sit alongside your system that is built up over time using precedent or it may over-rule it or simply replace it.

HOW CAN YOU MAKE IT EFFECTIVE?

Your legal system has to work. First you need a neutral forum – a courtroom perhaps – and to appoint a group of competent individuals to administer the system. Our equivalent might be lawyers, judges, and so on. These individuals need to be sufficiently knowledgeable and skilled enough to make a decision. They will have to manage the process and then sift through the substantial issues so that they can make the most appropriate decision. In order for their decisions to be respected they have to be seen to be impartial and independent. This could be a problem on your small island where everyone knows everyone else. Also, the Island Council may want to put pressure on the system by appointing someone who can act as their executive when enacting their policies.

The decisions you make in this forum have to be enforceable or they will lack teeth and will quickly lose the respect of your citizens. You realised earlier that different problems called for different solutions. Broadly this seems to divide into two or three categories.

The first involve problems that require punishment – what we would term a criminal system – where our moral code is broken, there is a victim who has suffered loss and the person found responsible is punished in an appropriate manner. Note that this punishment does not improve the life of the victim – other than giving them some modicum of satisfaction, perhaps. As a separate act the victim may get some compensation – stolen goods returned for example – but the punishment system does not compensate for the loss.

You also realise that the punishment has the potential to be a deterrent as well. In this way you will discourage others from crossing the line of what you determine as acceptable behaviour. Sometimes, though, the punishment might not be harsh enough to deter; at others timess the punishment may seem disproportionately harsh.

The second involves the problems that are effectively private disputes between individuals or groups. This is almost a private legal system. These disputes are very different. The moral code may not have been broken but there is a problem that needs to be resolved peacefully – and may possibly prevent a breach of the moral code should it escalate. (The phrase 'taking the law into your own hands' springs to mind.) Here an impartial decision may be all that is required. No breach of the moral code has occurred and a financial settlement may be appropriate to compensate for physical loss of property, for example. This private system can also resolve the disputes where a financial settlement is not appropriate – an agreement with my neighbour not to hold noisy parties through the night. If the decision is ignored I should be able to return to the forum for a method of enforcement.

The third group are problems that occur when the rules that the Island Council has made are broken. These may be neither criminal nor private but may upset the equilibrium of the Island community if they are not enforced. An example might be rules about the appropriate clothing for the beach or restrictions on when you can light a barbeque. These rules may change to reflect changes in attitudes too.

Lastly, you may want a right of appeal so that you can take the decision to another forum, possibly the Island Council. In most legal systems the decision-making is tiered and hierarchical: the lower courts, criminal and civil, make decisions which may be appealed. The decisions of the upper courts over-rule the lower courts and – more importantly bind all the lower courts to follow that decision in other cases. Often cases are sent to the upper courts because the lower courts do not know what to do and need direction.

But what happens if the dispute you have is with the Island Council itself or as a result of one of its actions? In most western democratic societies the executive arm of the law – the court system and the judiciary – are independent of the parliamentary system that makes or confirms the law. This is done in order to avoid political interference in individual cases although the parliament – your Island Council – may lead to changes in the law if it is not adequately reflecting the wishes of your civil society.

We are not going to be able to answer all of these questions. Exploring the problem of setting up your legal system has also left you with some possible different answers to some of the questions. The system relies upon a common moral code. Religious teaching may be a good starting point but what happens when there are multiple beliefs and multiple moral codes?

Also what is the role of the Island Council in making the rules, deciding the method of enforcement and establishing any rights of appeal? You are relying on the Council to show the same qualities of fairness and consistency and to act in a benign manner for the good of the community.

Broadly, though, your system will have a number of components that will not change:

- A moral code.
- A forum for hearing and making decisions.
- A Criminal system and a Civil system.
- A method of enforcement for both systems.
- A possible route for appeals to a higher authority.

The qualities you are looking for are consistency, appropriateness and well-explained impartial solutions that are arrived at independently.

It may come as no surprise that many of the features of your desert island legal system are present in the English Legal System – but not all of them. Hopefully this exercise, although simplistic, will have illustrated some of the problems that any legal system must deal with and will have helped to open your mind to the way any system must evolve and adapt to change. In the discussion a number of questions were not answered and a number of options were left. Different countries have chosen different options – in particular the role of the State in making laws and designing a system for hearing cases and enforcing them. You will also have seen the opportunities available for using and abusing the system to suit a particular political regime. Also, different systems have grown up independently or coexisted with other similar systems, internationally. Today these different systems have to work together in the global society and further systems have emerged to deal with the differences and potential conflicts on an international scale. In turn, other systems have impacted on our own system, European Community law for example – this will be discussed later in this chapter.

You may also have concluded that this commentary has been influenced by the author's experience, cultural background and knowledge. The same applies, of course to any design exercise. This reliance on your own reference points is as much a challenge in understanding a legal system as it is with architectural precedents. The English Legal System cannot be considered in a vacuum any more than your own design projects.

Also, on our desert island we have not developed the powers of our Island Council – or the enforcement system – but you will see that there is a strong link between political institutions, their social agenda and legal systems. You cannot study the legal system without considering the values the law reflects and supports. In following the commentary you will have developed a critical awareness of the key issues and begun to analyse the problems that face any legal system. Again, this is similar to the critical and analytical skills that you bring to your work as a designer. Fortunately we have side-stepped other more detailed procedural issues such as establishing proof and the role of evidence in making legal decisions as well as what happens when these conflict. However it is hoped that the exploration of the Desert Island legal system can be used as a useful starting point to investigate the English Legal System in more depth by considering the other topics in this chapter.

In concluding the exercise it may also be useful to look at the initial questions critically. The word 'justice' has also been deliberately avoided as an objective. That may be curious but the word is used in so many contexts that it is difficult to give it a single meaning. The desert island society must possess some level of order to allow citizens to relate to each other on a day-to-day basis and to resolve conflict when it occurs. A legal system has a central role in maintaining that social order, although it is not the only one. Religion, custom and a sense of moral purpose all have an important role to play as well. However legal order is a particular type of social order and has its own set of formal mechanisms of social control. As designers you need to be aware of the nature of that formal structure but understanding the principles about the English Legal System is more than recognising a set of legal rules, it is also about understanding a social institution of fundamental importance to our society.

THE CATEGORIES OF LAW

The English Legal System comprises a number of different categories that represent a development of the different strands identified on our mythical desert island. Unfortunately the terms commonly used often have a dual function which can be confusing. The different categories are as follows:

1. Common Law systems and Civil Law systems

This describes two distinct systems. England and Wales follow a 'Common Law' system. This is the historic English legal system and has been adopted by other countries such as the United States and many other Commonwealth countries such as Australia and Singapore.

'Civil Law' is the term given to the system adopted by most other European countries and has its origins in Roman law.

The general distinction is that the 'Common Law' was historically made by judges based on the accumulated experience of sitting hearing different cases. In this way a set of principles evolved in English courts over time. Civil law is based on a set of abstract principles that are codified according to different circumstances and which judges follow. In practice judges in a civil system have some discretion. The European Court of Justice follows a civil law system which leaves the potential for problems when translating European law in to English law and when cases move from England to the European system and then back again following a decision.

2. Common Law and Equity

These two terms represent an historical distinction between two distinct strands of the English Legal System whose foundations go back to the origins of the common law system. The distinction reveals a key characteristic of legal systems – the ability to adapt themselves to changing circumstances. First, it is worth expanding on the origin of the concept of Common Law, especially as it lays the foundations for many of the important systems throughout the world.

There is a romantic notion that the 'Common Law' represents the law of the common man spread by 'word of mouth' and is somehow a democratic system that starts at the base of society and moves to the top giving it both social acceptance and credibility. In fact, it describes a system that is common to the whole country. Before the Norman Conquest in 1066 there was no single legal system. Historians debate the impact of the Norman Invasion on the government of England. Recent interpretations suggest that the Normans inherited a relatively sophisticated and stable Anglo-Saxon legal and social system that grew out of the agricultural revolution of the ninth and tenth centuries. There was no need to alter this system until it broke down in the twelfth century and it was 1166 before a 'common' system emerged under the control of the King and a small number of judges were tasked with travelling the whole country sitting at various historic regional centres to hear cases and decide disputes. (No doubt it was adapted to achieve consensus amongst the major territorial landowners and fragmented institutions without whose co-operation it would not have been effective.) In effect the appointed judges were asserting the authority of the new state.

As time passed the system evolved and shifted from being an instrument of the King's rule. The common law courts took particular forms and worked to common points of reference and within relatively set parameters and formal procedures.

A flavour of the court system can be seen from an extract of surviving court records from the thirteenth century:

'It is presented that Alice wife of Baugecler broke into the barn of Robert son of Henry de Crigeliston and stole four sheaves of corn. Robert pursued her and took the corn from her, and allowed her to go. Fine 13 shillings and 4d (67p).

Agnes Lambot against Adam son of Richard de Alvirthorpe, says that he beat her, which he denies and they put themselves on an inquisition....he says that he did beat her because she used insulting words to his wife. He is amerced (fined) 12d (5p) for using insulting words to the plaintiff and the plaintiff is amerced the like to Adam's wife.

Isolda de St Oswald against Jack de Ireland, who was seized and imprisoned for larceny (theft) says that on the Sunday before Christmas, in the night, he stole a robe of burrell, trimmed with black lamb-skin, value 8s 6d (43p) which was in his keeping. He came in full Court, before a Steward and before John de Horbiry, the King's Coroner, and all the suitors of the Court, and confessed it with his own lips. Therefore let him be hanged.'[45]

In this short extract we can see a range of disputes from theft to a neighbour dispute that ended in violence. It appears that in this case the fine was for the insulting words rather than the beating.

However the system as it developed failed to deal with matters that apparently fell outside the system. This created the scope for injustices. If you had the resources you appealed directly to the sovereign for a decision in these cases and decisions were delegated to his Lord Chancellor who would act as 'the King's conscience'. As the common law courts became more formalised these pleas increased. The response was to develop another system, the Courts of Equity which were set up to deliver 'fair' or 'equitable' decisions in cases that the common law court system failed to deal with. This system also had its problems; for example, it was essential to bring an action in the right court. These two systems ran in parallel until they were unified under the Judicature Acts in the late nineteenth century. Interestingly, these changes led to the demand for a new type of building. The resulting architectural competition for new law courts was won by George Edmund Street – his famous Victorian Gothic-style Royal Courts of Justice setting for many equally famous cases today.

COMMON LAW AND STATUTE LAW

You will have seen that the common law started as an instrument of the sovereign but developed into a relatively autonomous system. The legal core of the common law developed from the cases heard and decisions that were made over time – precedent, used in much the same way that architects use precedent in developing building designs – together with a set of procedural rules. This process still has an important role in developing the law to suit new circumstances.

In our desert island example you saw that there was a need for the Island Council to develop a set of rules rather than let them evolve. It is to be expected that the same applies in a complex democratic civil society such as the UK. Statute Law refers to the new laws – or legislation – created by Parliament. Although the growth of statute law has been significant and affects almost every aspect of our daily lives the courts are relied upon to determine the operation of legislation.

Parliament creates statutes in a number of ways: it can react to events and introduce legislation to deal with particular problems as they arise; or it can be proactive – using statute to drive particular policies or principles. Lastly it can step in when the common law system is seen to be failing to deal with new circumstances or a change in society's values.

One example of Parliament *reacting* to events is the way it dealt with the perceived terrorist threat post-9/11. One solution was to arrest and detain suspected terrorists in order to prevent similar tragic events occurring. As a result the government lengthened the period suspects could be detained without charge and a range of periods were discussed and implemented under emergency legislation which would have been inconceivable only months earlier. Meanwhile the USA flew suspected terrorists outside normal jurisdictions so that they could be detained indefinitely without charge.

An example of a *proactive* stance is the Human Rights Act, 1998. This was promised in the Labour Party manifesto that led to their election to power in 1997. The Act adopted the international human rights legislation that governed the European Court of Human Rights and applied it to the English courts. Ironically it was the same government that extended the period that a suspect terrorist could be held without trial.

More often, Parliament combines the two approaches and decides – often for political reasons – to amend existing legislation in response to pressure that might otherwise be ignored or at least not be prioritised.

Increasingly Statute Law has stepped in to amend common law principles – either when it is seen to be failing and/or not reflecting changes in society. For example, traditionally, the contracts that are made between individuals and between individuals and organisations have been seen as 'private law' and Parliament has been reluctant to interfere. (You will see this again in the next chapter on contract law.) However one area of 'private law' – employment law – has seen a huge growth in legislation. This is primarily for three reasons: first, the inequality of bargaining power between worker and employer was out of balance and seen to be in favour of the employer. In short, the opportunity to bargain freely was not working. Second, the growth of worker representation changed society's attitude to this imbalance; and thirdly, Parliament – and in particular Labour governments (supported by the trade unions) acted proactively based on agreed policies to introduce legislation that adjusted the bargaining balance. The result has been a huge increase in employment law. It has now extended beyond simple contract terms and rights to reflect changing attitudes to sex discrimination and rights of parents to take maternity and paternity leave.

Employment legislation affects us all – as employees and employers. In the built environment there are specific areas that have seen a growth of legislation. These include:

- Town planning – where common law rights have been superseded by legislation that affects the urban and rural environment.

- Historic Building legislation that seeks to preserve the urban environment.

- Environmental legislation to protect the countryside and the quality of water and air.

- Sustainability legislation, which controls building methods through detailed regulations.

- Health and safety legislation aimed at making construction sites safer places to work.

- Equality legislation that makes spaces and places accessible to all.

These subjects will be explored in more detail in later chapters but there is a further area that affects both the common and private law and public statute law – international law and European law in particular.

INTERNATIONAL AND EUROPEAN LAW

We live in a global environment, and no more so than in the construction industry. For example, an English architect may be designing an airport in the Far East that is built by a French construction company with steel from India, glazing from Switzerland, built by local labour in accordance with a combination of America and European building legislation – for an international client financed by banks in New York and China. The architect's design team may include many different nationalities and the wider design team may include multi-national engineering and other professional consultancies.

However international trade and co-operation at a political and social level requires operating with multiple legal systems. Even if you agree a set of rules how do you enforce them across borders outside the limits of your jurisdiction? This is not a new problem but as we increasingly rely on international trade and even the most basic of commodities are sourced from many different countries it becomes increasingly important that we develop mechanisms to allow trade and also meet different sets of legislation.

Different countries – using their parliaments or political fora to implement policy – may use their powers to pass legislation that erects barriers to international trade in order to protect domestic markets. This can be done cheaply and effectively by charging tariffs or taxes on imports, for example, or setting specific regulations that imports must meet – again at little cost to the home nation but inevitably adding cost to the importer.

Over time, multi-lateral and bilateral international agreements have developed in many areas, mainly in trade. Naturally there is a tendency to harmonise trading laws with the neighbours you do the most business with. But these agreements are complex and – by their nature – require consensus. Also, they may be affected by the sort of imbalance mentioned above in connection with employment that will benefit one country over another. However international agreements *have* succeeded in opening up markets – even where they operate in favour of one country over another. With this background in mind we can consider European Law.

European Law – like the origins of English common law – can be seen as a political initiative, part of what has been called the 'European Project', the European Union. The origins of European law can be traced back to the efforts of European countries to recover from the Second World War. Progressive co-operation between countries – first in the trading of coal and steel led to the wider European Economic Community and the removal of trade barriers between member states. A European Council of Ministers and Parliament was also set up to harmonise laws that were perceived as barriers to free trade within the EU and to apply those laws to member states. The United Kingdom joined in 1971. The main advantage was the free flow of trade between members. However membership also meant that the UK, for the first time, would be subject to European law. (Note this is different from the European Court of Human Rights mentioned above which sits outside the EU.)

There were a number of problems to overcome. First, other European states operate under different systems governed by Codes laid down by their parliaments, rather than the English common law system that has evolved over centuries. The relationship between the courts and government is therefore different and the way the judiciary operates is different. Second, for centuries only Parliament had the power to create laws – this is called parliamentary sovereignty. It is therefore not possible in the English legal system to automatically adopt a law from another country. As part of the EU membership process we have to incorporate EU laws and Directives into our own legislation. If we fail to do so – and some member states do – then we can be subjected to sanctions. In practice EU laws are reinterpreted to suit local conditions provided the key legislation is met. The objective is to harmonise the movement of people, goods and services across the member states. Inevitably this has expanded into other areas.

The main way European law is enacted in the UK is through European Directives. We 'adopt' Directives by either creating a new law or sometimes by placing the Directive within an existing piece of legislation. For example, in the area of Health and Safety the European Directive sits within the existing Health and Safety at Work Act (HSAW) 1971. This avoids the need for Parliamentary time to be taken up drafting a new Bill and for Parliament to debate and pass it. The CDM Regulations 2010 – governing safety on construction sites – sits within the HSAW Act 1971 even though the Act predates the European Directive.

An example of how different countries apply Directives can be seen with the Working Time Directive 2003 (which itself amended a previous set of rules). In the UK the amended Working Time Regulations allowed workers to work up to 48 hours in a week. In France the limit is 35 hours a week.

CONCLUSIONS

The law represents a set of enforceable rules that are set down by society. As such it is subject to moral, social, political, international and more recently, economic influences.

The law has a number of key functions: to punish 'wrongs' – crimes such as theft and fraud and to enforce individual rights set out by a moral society.

Its courts also provide a forum for settling private disputes. A judge acts as a neutral third party, makes binding decisions and provides the means to enforce those decisions.

The English Legal System is made up of two key parts: the common law that is 'judge-made' and has grown up through custom and legal precedent and statute law – the rules created by Parliament.

Parliament uses the legal system proactively to implement policy and reactively by responding to events or changes in society. It also steps in when the common law appears to be failing, in employment law for example.

Only Parliament can create new laws but increasingly English law is affected by international treaties and agreements and European Law created by the European Union Council of Ministers and European Parliament.

Laws made in Parliament may feed down to society by the creation of sets of Regulations, such as the Working Time Regulations 2003 or CDM Regulations 2010. Although Parliament depends on the courts to operate legislation and settle disputes the regulations created by Parliament are generally administered by government departments, such as the HSE in the case of the CDM Regulations or local authorities in the case of planning regulations.

In the next chapter you will see how 'private law' operates in the creation of formal contractual relationships and how we are also subject to other laws that seek to preserve our rights in a civil society.

FURTHER READING

Martin J 'The English Legal System' Hodder 2010

Slapper G 'The English Legal System' (3rd ed.) Routledge 2012

Wevill J 'Law in Practice' RIBA Publishing 2012

6
LEGAL RELATIONSHIPS –

HOW WE ESTABLISH DUTIES, RESPONSIBILITIES AND OBLIGATIONS

INTRODUCTION

The law used to resolve private disputes – private law – is the main area of the legal system that affects the day-to-day lives of design professionals. This is important because the deals that you, i.e. architects and designers, agree with your client set up a series of obligations: to devote your time, knowledge and design skills to help your clients to achieve their objectives; your clients in return contribute their time, give you feedback, and – normally – pay for your services. This is known as the client-architect relationship. Typically, this is set down in writing as a legal contract, a process controlled by the part of English common law system that is called, unsurprisingly, the law of **obligations**.

Another part of the law of obligations is the law of **torts**, meaning 'wrongful acts'. It enforces your wider duties to respect the rights of your fellow citizens – duties that cannot be set down in a contract. Because the law of torts is potentially far-reaching and very different in character to the rules governing contracts, it has its own set of rules, developed by precedent, that imposes greater duties on professionals than it does on other citizens.

The law of obligations also includes the principle of **restitution**, meaning 'compensation'. For example, if you incur huge expense, time and effort preparing a feasibility study on the understanding that you will end up with a contract – but it fails to happen the law will tend to be unsympathetic: it is your own fault for being so foolish as to work at risk. However, there are circumstances where it will agree that you should be compensated. Usually, though, it applies where someone has mistakenly or otherwise acquired, for example, property or money without justification, and the law requires it to be returned to its rightful owner. In other words, there is no formal contract but the law steps in anyway to achieve a fair result.

The public court system provides a neutral forum to make decisions that enforce these rights, provide restitution and to make compensation.

There is a general principle that the agreements or contracts between individuals and/or private companies should remain private. The state is reluctant to step in unless private law is seen to be working either inequitably or unfairly, or it has failed to keep up with societal changes. When that happens, Parliament imposes new rules enforced through Acts of Parliament. The most prominent example of this is in employment law, where the fundamental inequality of bargaining power between employer and employee led successive governments to legislate new rules and procedures.

Being aware of the law of obligations in the architect-client relationship and in how you practice architecture is important. First, private contract law defines the rules that create formal relationships, which of course you must adhere to. Second, the law of torts imposes on you a series of wider obligations to society. Third, restitution might be an appropriate instrument of rescue for unfair treatment not covered by a contract. Finally, the state, through Acts of Parliament, periodically introduces new rules which you have to keep an eye on.

The rest of this chapter will concentrate on contracts and some of the wider wrongs that come under the law of torts but may directly affect architects and designers.

WHAT MAKES A CONTRACT?

Because a contract is enforceable, you need to know when you have one and when you don't. If it's possible to say that you don't have one, a party can wriggle out of his or her obligations. Luckily, there are common-sense rules that help us to decide when a contract exists.

We establish many relationships and agreements on a day-to-day basis. Not all of them have the status of a contract and it is unlikely that we would want them to be enforceable anyway. For example, you might want to borrow a friend's bike because your own has a puncture. You say that you will buy them a drink when you get back. He or she agrees but would like it back by late afternoon as they are going out in the evening. Instinctively you agree and a bargain is struck. It is unlikely that your friend will want to be compensated if you are inadvertently late: your offer of a drink should be enough to keep the relationship sweet.

It is a distinct advantage to keep agreements like these informal – if we didn't, we would be weighed down by formal contracts… and have no friends. However, failure to meet your side of the bargain or keep your promise is the seed of disagreement, which, depending on the circumstances, can grow into something nasty.

Interestingly, there is no formal definition of a contract in English law. Even major authorities on the law are tentative on the matter:

'The law of contract may be provisionally described as that branch of the law which determines the circumstances in which a promise shall be legally binding on the person making it.'[46]

The lack of a definition can be attributed to the fact that contract law did not develop from an abstract theory but evolved organically in response to disputes brought before the courts. Until the common law was reformed in the nineteenth century, the courts concentrated on the procedures rather than the substance or content of any claim. Only later did legal experts set out the general principles of a law of contract.

Although there is no agreed definition, the principles can be set out with reasonable certainty. There must be:

1. An *agreement*.[47] Usually, you must both promise to a set of actions.

2. *Consideration* – something of legal value in return for the promise. This usually takes the form of money paid in return for effort or delivery of a product.

3. The *intention* to create a legal relationship. What is in the mind of each party as well as their behaviour is relevant.

You can test the bike example against these three principles. Yes, there was an agreement – loan of the bike for a fixed period of time. There was consideration too: you agreed to buy them a drink. However, there was no intention to create a legally binding relationship that could be enforced. Even if you were late and did not keep your side of the bargain, commonsense says that it would not go any further. So, the *intentions* of those who enter into the contract are important.

Note also that there is an implied process – a sequence of phases – taking place.

THE DIFFERENT PHASES TO A CONTRACT

The phases of a contract are what you'd expect, and almost all contracts go through them.

1. Before the contract.

2. Forming the contract.

3. Performing the contract.

4. Completing the contract.

Each phase has potential hazards waiting to trip you up. A contract is a promise or bargain based on a mutual understanding between the parties. Unfortunately, parties rarely share the same perspective or attitude, making it all too easy for their intentions to differ.

Take the bike example again. On reflection you aren't sure you can return it on time but aren't too worried because standing a round at the pub will sort it out.

However, your friend assumes you understand that the return time is critical, and the promised drink, nice offer though it is, will in no way compensate for being late for her night out. Same scenario, but now it's clear the parties have very different expectations, and it's a dispute in the making. Luckily you don't have a contract.

BEFORE THE CONTRACT

In the construction industry, a lot of effort goes into the period before a contract is formed. Architecture isn't as simple a transaction as, for example, buying a laptop computer. Our professional relationships are generally long-term and built up over time. Our obligations are not always clear-cut: the scope may change, and projects may be transformed over the project lifecycle. In many cases clients look to their designer to help define the project and determine the professional design services that need to be provided. In short, our design services at the beginning of a project are rarely clearly defined and the architect-client relationship is often informal.

This is a dangerous time, and the architect-client relationship, if it is to flourish, needs to be put on a formal footing. Some designers are reluctant to do so, however, fearing that moving to a formal contract will somehow upset the relationship or restrict their freedom. It is bad practice for a designer to rely on what he or she believes a client was thinking at a particular time even when the relationship is good. If you fall out, though, the last thing that will happen is that you retrospectively agree on the fundamentals. For this reason alone it is important that architects set out their services clearly and in writing as soon as possible.

In the desire to get appointed it can be tempting to misrepresent what you can do or achieve. There are a number of dangers here. You need to be clear that you can deliver what you say you can deliver – you must have the necessary resources as well as the expertise. Also, clients often believe that architects have more control than is the case. For example, they might think you can guarantee planning permission – and if you don't then you have failed. Of course, architects have no such power – planning decisions are decided by local authorities. Never hold out that you can do something that is beyond your competence or control.

FORMING THE CONTRACT

For a legally-binding contract to be formed you have to show that the three key principles – agreement, consideration and intention – have been met.

It sounds simple but can be complex and lead to disagreement. Remember: the aim of this process is clarity and certainty. Where the offer and acceptance are unclear and lack certainty, there is fertile ground for misunderstandings and dispute – even before the contract has started. This is especially true in the construction industry due to the complexity of the services provided and the 'one-off' nature of the product. Again – very different from walking into a store and buying a laptop.

The first principle is met by showing that an *offer* has been made by party A and that it has been *accepted* by party B. An offer is a definite statement that you are willing to enter into a contract. It is not the same as an invitation to negotiate, which may or may not lead to an offer. An acceptance is a final acceptance of an offer.

For example, it is now common for clients to ask architects to quote for their design services in a process called tendering. (This is traditionally an opportunity for unknown but talented designers to compete with more established practices.) The client's aim is to achieve the best value, service and end-product by making the process competitive and transparent (i.e. fair). In common law, an invitation to tender for project work is an invitation to negotiate and *not* an offer. The designer's quote or tender, however, *is* an offer, and the client can accept it or not. The tender process does not therefore bind the client to accept any offer. Once accepted, though, the offer is binding. This is because both intended to create an agreement that is enforceable in law.

Note that you don't need a signature for there to be a contract – an oral agreement can be binding, provided it meets the three principles. However, this is risky and unprofessional.

Formula 1 entrepreneur Eddie Jordan launched a claim in London's High Court saying Vodafone wrongly pulled out of a three-year deal to sponsor Jordan's cars and went on to back the rival team Ferrari.

"You've got the deal," were the words Jordan claimed were spoken to him on the telephone by Vodafone's global branding director David Haines.

Jordan claimed these words sealed the agreement for the three-year sponsorship of its F1 cars on the terms negotiated between the parties in the prior months, even though no written contract was produced.

Vodafone argued that it merely entered into negotiations with Jordan, along with rival racing teams McLaren, Benetton, Ferrari and Toyota, as part of its global branding strategy.

Mr Justice Langley formally handed down the highly critical judgment leaving Mr Jordan's reputation in tatters and his Formula One team facing a potential legal bill of about £5m.[48]

Dismissing the case the judge said that Jordan's £150m lawsuit against Vodafone was "without foundation and false" as well as "contrived and unsustainable". And that "the inherent improbability of an agreement of such a nature for payments of such a size being made in such a manner is obvious".

In common law, the point at which the contract is formed is important. The terms that are understood on agreement will be a point of reference, especially if there are subsequent changes – as happens all the time in design and construction. Where there is so much scope for uncertainty, you must agree the following:

1. The names of the parties – you and your client.

2. The scope of services – what you are going to do.

3. The cost of the services – or how you are to be paid and when.

4. Who will be responsible for what – this includes any limitations on your services.

Chapter 2 discussed the high standards expected from professionals, standards that sometimes exceed normal commercial requirements. Because the architect-client relationship is so important and contracts are your lifeblood, the profession sets high standards for our agreements. Both the RIBA in its 'Code of Professional Conduct' and the ARB's 'Architect's Code' set out best practice for architects' agreements. For example, while an oral agreement can be legally binding, the profession requires all professional agreements to be in writing. To help, the RIBA publishes a set of standard forms of agreement that are routinely updated in consultation with major stakeholders, clients and representatives from the construction industry.

The construction industry has also developed model procedures for tendering and a set of standard contracts for construction projects. Again, these contracts are routinely reviewed and amended to reflect changes in industry practice and case law.

PERFORMING THE CONTRACT

A legally binding contract needs consideration, which means 'something of value' or benefit, usually fees in return for effort. You *perform* design or consultancy services in return for payment.[49] The agreement usually sets conditions on this: you need to be competent, have the requisite resources, and demonstrate the necessary skills and knowledge. Equally, your client needs to have the funds available to pay you at agreed intervals or at the end of the contract.

At the heart of the contract, therefore, you need to know what you are performing, what are you going to be paid, and when.

The only way to record what you are performing, even on simple jobs, is to write it down. A 'gentleman's agreement' will almost certainly lead to misunderstandings, and is too risky when your professional life depends upon it.

It is easier to agree your fee, especially if the payment is based around a single event such as delivering a design. However, even this can get tricky. Delivery may take months – or years – and in the meantime you need to pay yourself, your staff, your landlord and the utilities every month or every three months. A single payment event is likely to be unviable, and you will almost certainly need to agree regular (monthly) part-payments. You won't be too popular with your bank otherwise.

Design and construction projects tend to follow a recognisable process/lifecycle, and architects tend to carry out the similar professional services along the way. As a result the profession has developed a standardised approach to both – the RIBA Outline Plan of Work (OPoW). The OPoW is useful in that the design team and construction professionals readily understand its stages and the work expected of them at each. The RIBA's standard professional contracts are based on the key work stages in the OPoW.

The list of services and their sequence follows a simple cycle. Payment terms may be linked to particular stages or can be charged on a regular time basis. Both parties to the contract can see what their obligations are and when they are to be performed. For example, completion of an outline design at RIBA Work Stage 3 'Developed Design' (OPoW-2013) may trigger a specified payment set out in a separate short schedule of payments (But remember also what practitioners said about payment terms in chapter 3).

COMPLETING THE CONTRACT

The contract is completed when the parties have performed or *discharged* their obligations completely. It is important to reach a point where all the obligations have been met. The guiding principle is that *all* the requirements *must* be met – performance and payment. Note that if these terms are not set out clearly then it is almost impossible to know when this point is. And if you can't tell, you might struggle to be paid in full.

This demonstrates how the key principles of a contract link together, and emphasizes the need for professional services contracts to be in writing.

TERMINATION

A good, legally binding and enforceable contract cannot be broken easily. In other words, it's serious: you cannot walk away from it just because you have changed your mind. Neither can you break the contract if you have a mere disagreement – it has to be a serious and substantial breach of the terms. If you decide unilaterally to walk away, you will still be responsible for the performance of your part of the agreement. The *burden* and the *benefit* still exist. If this is not practical, professional services contracts must contain some mechanism for ending the contract prematurely – this is called 'termination'. To avoid further disagreement, termination is triggered by a defined event or events agreed in the contract. Because termination is a major event in itself and there must be no ambiguity, the mechanism includes a protocol for formal notifications. Remember, one of the central aims of a written contract is 'certainty'. If this fundamental event is not handled correctly then the actions (or inactions) that flow from it can result in another disagreement and a possible claim for a breach of the contract terms.

PRIVITY OF CONTRACT

Contract law is guided by another principle: the promise or bargain you make is exclusive to those that make it. The legal term is **privity of contract.** This means that the burden (for example, your design services) or benefit (the money you receive) cannot apply to anyone else. The legal consequences are that the contract cannot be enforced by or against anyone who is not a party to the contract. This is important because it defines the limits of your obligations and creates certainty. It does, however, create problems in the complex multi-party environment that characterises the construction industry, an issue that is beyond the scope of this book.

THE GOVERNMENT STEPS IN

Successive governments have held back from interfering with our freedom to make promises and strike agreements that are legally binding. On occasion, however, they have had to. (For example, employment law has grown in recent years to address the imbalance in bargaining power between employer and employee.) The key legislation is discussed below. It is not intended to be comprehensive, and covers key points only.

The *Misrepresentation Act 1967*

Changes in public policy have led to the introduction of new laws that affect contracts. This is partly due to the ascendency of the consumer in society. We are all consumers and need some guidance or safeguards when entering into contracts. Although the statutes affecting contracts apply to all citizens and organisations, the consumer is generally given greater rights. This is to reflect the fact that a consumer is not an expert, a point that is particularly important for contracts with owner-occupiers.

The pre-contract period has always been a problem area as consumers, in particular, can be vulnerable because of an inequality of knowledge – the provider knows more that the consumer. A statement relating to a contract but which does not become part of the contract – a *representation* – should therefore be accurate. Information of this sort that is inaccurate is therefore a *misrepresentation*. Government reacted with the *Misrepresentation Act 1967*, which provides safeguards and remedies against this kind of thing. The effect was that information provided during the pre-contract period became much clearer because (both) parties ran the risk of being prosecuted for making inaccurate statements – even innocently.

The *Sale of Goods Act 1979* and the *Supply of Goods and Services Act 1982*

The *Sale of Goods Act 1979* governs contracts for the sale of goods. Its most important parts set down a series of terms which are to be *implied* in all relevant contracts. In other words, these terms need not be expressly written into the contract to be effective. Where the goods are described in the contract – which is usual in construction work – the quality of the goods must comply with the description. There is also an implied term that the goods will be fit for purpose. These principles were extended by the Supply of Goods and Services Act 1982 to include services such as design services. The common law position under a contract is that architects should carry out their services with reasonable skill and care.

Note that it is the *contractor's* responsibility to meet the implied term that your design is built to be fit for its intended purpose. Designers cannot guarantee this as they do not contract to build the object ('supply the goods'). This is an important distinction when the line between design services and construction are blurred as is increasingly the case in modern, complex construction projects. For example, if – unusually – you contract to design *and supply* the built form then it is implied both that you will design with reasonable skill and care and that the object will be fit for purpose.

The *Limitation Act 1980*

The last important piece of legislation, the *Limitation Act 1980,* defines how long you are liable for your services under a contract. In fact, it limits liability to either 6 or 12 years. For design services and construction contracts, the time period begins from the date the works are completed. This may be different from the time the contract duties – such as final payment – are completed.[50]

THE PROFESSION STEPS IN

Because contracts are the lifeblood of the profession in practice – we want to use our professional skills and we want to be paid for them – the profession has developed extra standards for our professional agreements with clients. These are more onerous than the requirements under common law. Both the RIBA and ARB set out requirements in the Code of Professional Conduct (2005) and The Architects Code: Standards of Conduct and Practice (2010) respectively and give detailed guidance on their websites. A failure to follow these requirements can give rise to disciplinary action even if you have a contract recognised in common law. For example, an oral agreement to carry out any professional service as an architect is not enough – it has to be in writing. A snapshot of the relevant Standards is given below.

RIBA Code of Professional Conduct 2005

'Principle 2 – Competence

2.3 Members should ensure that their terms of appointment, the scope of their work and the essential project requirements are clear and recorded in writing. They should explain to their clients the implications of any conditions of engagement and how their fees are to be calculated and charged.'

ARB Architect's Code: Standards of Conduct and Practice 2010

Standard 4
Competent management of your business

1.4 'You are expected to ensure that before you undertake any professional work you have entered into a written agreement with the client which adequately covers:

- the contracting parties;
- the scope of the work;
- the fee or method of calculating it;
- who will be responsible for what;
- any constraints or limitations on the responsibilities of the parties;
- the provisions for suspension or termination of the agreement;
- a statement that you have adequate and appropriate insurance cover as specified by the Board;
- your complaints-handling procedure...'

WHAT HAPPENS IF IT GOES WRONG?

When you fail to meet a contractual obligation, the agreement is broken. This is described as a *breach of contract*. Because your agreement is legally binding, you or your client can apply to the courts to obtain a remedy or *damages*. The courts do not have the resources to compel a party to perform the services that they agreed to carry out and so will generally award a financial remedy where this is appropriate. The principle in contract law is that the damages awarded will take you to the position you would have been in if the contract had been fully performed.

> For example, if you have not been paid £3,000 as agreed for a house design that you have completed, your client's failure to pay is a breach of contract. You can go to court to obtain payment. You would expect to be paid the full £3,000. In principle, you can also claim the costs that are directly associated with the breach, including taking him to court. The extra costs act as an incentive to settle promptly.
>
> Alternatively, if you failed to complete a design and it was only half-finished and six months late then your client may argue that you have broken the terms of the contract. If he goes to court, the court cannot force you to complete the work but he can claim the cost of taking the contract to the point where it will be completed. This may exceed the original agreed value of the work. The same principle concerning costs also apply.

As design is a complex service and the product is generally a 'one-off', disputes that end in court are much more complex than the simple example given (above). Most disputes occur because clients do not receive the services they expected at the time they expected them because the scope of services was badly defined. In case it needed emphasizing, a vague contract leads to a greater risk of a dispute: best to always use a standard form such as those published by the RIBA.

Another scenario is where a difference of opinion gives rise to a claim and counter-claim. For example, your client has not paid his invoice for some sketch designs for a new house within the period stated in your contract. You want payment. He says that you have not delivered the agreed design in the agreed timescale, and anyway he is not very happy with the standard of the work. He wants to see a more highly-developed design. Both of you are making a claim that there has been a breach of the terms of the contract, and, because you disagree, there is now a dispute.

A breach of contract has two main consequences. First, the breach gives rise to a right to claim damages. Secondly, if the breach is sufficiently serious and goes to the heart of the contract, it may be possible to bring the contract to an end. The point at which the contract may be terminated is difficult to establish and beyond the scope of this book. The two claimed breaches in the example above, however, are not sufficiently serious to terminate the contract.

The remedy for your claim is for your invoice to be paid. The remedy for your client is to receive the service and the standard of drawings he was expecting, and if that is not agreed then the remedy will be financial. The award would take them to where the contract says they should be – and reflect the cost of the complete set of drawings.

The vast majority of claims and disputes are settled before they reach court. It could be argued that the public courts are not the place to resolve private contractual disputes. The costs alone of preparing a case for court are often enough to encourage the parties to reach a settlement.

ALTERNATIVE DISPUTE RESOLUTION

Increasingly, disputes are resolved using other methods and places. For example, most cases centre on factual disagreements and not complex points of law that need to be argued and decided by a highly-qualified judge. A much better route is to negotiate a settlement, which usually means sometimes significant compromise on both sides. **Mediation** – which is now actively promoted by the courts – involves bringing in a neutral person to act as a facilitator. The dispute remains in the hands of the parties (in contrast to the court system where decision-making is passed to the judge) and the mediator merely assists the parties to arrive at a settlement. It is also kept private. **Arbitration** is another method: an arbitrator is appointed by both parties to hear the case and make a decision. It can still be costly but the process remains under the control of the parties, and again is private.

The construction industry is traditionally fertile ground for disputes. A review of the construction industry in the 1990s, the Latham Report, recommended a new way of resolving disputes quickly and inexpensively – **Adjudication**. Unusually, this was adopted by the government, and statutory adjudication is available as of right to parties to construction contracts – this includes our professional contracts with our clients. The process has a time-limit and the adjudicator's decision is binding – in other words it can be enforced by a court if necessary. The majority of construction disputes including contracts for architectural design services are resolved this way, and the statutory process has created its own micro-legal system. However, as the process has matured it has become increasingly legalistic, and moved away from the 'rough justice' of its early years. With increased sophistication has come increased cost. Parties are now seeking more cost-effective methods such as mediation or other methods that try and settle claims before they crystallise as disputes.

OBLIGATIONS TO OTHERS – THE LAW OF TORT

The law of obligations also extends beyond contractual agreements to an obligation to other citizens, covered by the law of tort.[51] This allows a person who has been wronged to obtain damages other than for wrongs caused by a breach of contract.

In the context of architecture and design, the law is mainly concerned with providing compensation for personal injury and damage to property caused by *negligence*. There are however a number of other categories of tort law.

These include *trespass* and *nuisance*, which both apply to the protection of land and property rights. It also protects other interests, such as your reputation – *defamation*, and personal freedom – *assault* and *false imprisonment*. The fact that it deals with rights, physical damage and harm means that Parliament has also bolstered it with legislation. However, the tort of *negligence* is no more significantly affected by statutes other than the area of compensation for injury – which suggests that it works sufficiently well for the needs of society. The scope of tort law is very wide but it is mainly negligence that architects need to be wary of.

NEGLIGENCE

Negligence can be defined as carelessness that amounts to the breach of a duty – a failure to do what a citizen could reasonably be expected to do.[52] As professionals are recognised for their particular knowledge, skills or competence, tort law expects the standard a professional architect to meet is higher than the average citizen's – but no higher than an average member of the profession's. This is a very important principle as it defines a reasonable standard of professional competence.

One of the purposes of architecture is to create or improve the built environment. We build for the wider public – not just our immediate clients. We celebrate the wider effects of architecture on a range of stakeholders and society. However, privity of contract limits the reach of the law to the parties to the contract: A and B. What about duties to other stakeholders C through to Z? The problem is shown in the following true event taken from a local newspaper.

'Chickens crushed to death after hot air balloon panic: farmer gets cash compensation from Virgin

'Nearly 150 chickens were found dead after a hot air balloon scared them to death. It is thought that the birds had climbed on top of one another in a desperate attempt to escape the large red Virgin balloon which they had seen from a shed window. "I opened the shed to collect their eggs and there they were."

'After the chickens were discovered, farmer Steve Collins contacted Virgin Balloon Flights who have since offered him compensation and a balloon flight. The pilot has also visited the farm to apologise. A spokesman from Virgin said that in his 25 years of flying he had not known chickens react like this, but he is very sorry.'

The Bucks Herald, 18th April 2012

From the example, you can see the legal problem – the farmer has suffered a loss but does not have a contract with Virgin Balloons. He cannot make a contractual claim but he can make a claim in tort. Luckily, Virgin has accepted its liability in tort and offered compensation – and a free flight. The example, although not architectural, demonstrates how the tort of negligence works. If Virgin had denied liability and the farmer had made a claim, he would have had to prove a number of key points:

1. Was the pilot negligent? In other words, was he careless or as a professional did his standard of flying fall below that of an average hot air balloon pilot?

2. Was there a close link between the cause and the effect?

3. Did the negligent act (the balloon flight) result in physical damage or harm?

You do not have enough information to answer the first question – and this is the most contentious. But if it could be shown that he was flying too low or ignored instructions then that might be sufficient to prove negligence – in comparison with the standard of an average pilot.

The answers to question two is 'Yes.' There is a definite link.

The answer to the third question is also 'Yes.' The chickens died.

The farmer could claim damages, which is the legal remedy in tort. The damages in tort are different from the damages in contract. Whereas damages in contract take you to the position where you would have been if the contract was performed in full (in other words, a point in the future), the damages in tort compensate you for the loss – they take you to where you were before you suffered the loss (a point in the past). In this case, compensation would be the cost of replacing the dead chickens. In this instance, Virgin immediately offered compensation and threw in the offer of a free flight to help maintain good relationships and generate good PR. (Note – we do not know if Virgin admitted liability in tort but it is unlikely.)

The following example shows how the principles apply more specifically to construction and property.

> You designed a small apartment building for a developer client, AA developers, who were enthusiastic about your approach to innovative design. Your agreement for the design was therefore with AA. You carried out everything you were asked to do and AA paid you in full. In fact, AA was so happy with the design that it recommended you for other work. Then AA sells the apartments and after a year the roof starts to leak and damages the flat on the top floor. Furthermore, the leak damages the contents of the flat, including an expensive music system. And now the owner of the flat wants to claim for the damage. She calls in an expert who says that the work is satisfactory but the design was at fault.

She cannot claim against you because her contract is with the developer (the principle of 'privity of contract'). She *can* claim against the developer but she can also claim against you under the law of tort, specifically negligence. To make a successful claim she has to prove three things:

1. **Did you owe her a duty of care – to design the roof with 'reasonable skill and care'?**

   Answer: it is reasonable to expect that as an architect you designed the building to be occupied. She occupies the top flat therefore you owe her a duty of care in the tort law of negligence.

2. **Was there a breach of that duty?**

   Answer: she has to demonstrate a breach of that duty to design with reasonable skill and care. This is a high barrier and higher than your simple contractual duty to your client but there are similarities, for example you contracted to design with reasonable skill and care. Even if this is not mentioned in your written contract, you know that it is an implied term.

   She also has to prove that your work fell below the standard of another qualified architect. (Note that if you said you were an expert roof designer you will be measured against similar experts.) For example, if it could be shown that you had not followed published design guidelines or had not kept up-to-date with new legislation then you may have fallen below the standard required.

3. **Did the breach lead to the damage?**

   Answer: Even if the answer to the first two questions is yes, she still has to prove that the failure of the design caused the water damage. In this case it would depend on the facts. It is very likely that a failure in the roof design would cause a leak and the leak could damage the room(s) in the apartment below.

The question of the other losses that flow from your negligence, such as the expensive music system, is beyond the scope of this chapter.

The damages due to the flat owner should be an award that takes the condition of the apartment back to where it was before the damage occurred.

What about the developer? She could make a claim against the developer under the contract to purchase the apartment, and the developer would then make an identical contractual claim against you. In fact, as these kinds of issues are predictable, the developer is likely to have insisted on you signing a *collateral agreement*.[53] This is a simple way to establish a contractual link between the purchaser and you, the architect. The developer can ask you to enter into a contract with his future purchasers that mirrors the contract that you have with him and runs alongside your own contract. It's great for the owner: for her claim to be successful she only has to show a breach of contract. It's also great for the developer because he remains outside the contractual loop. Sadly, you are unlikely to be paid more than a token sum for performing this contract.

SUMMARY

In this chapter you have seen how our obligations to each other and to society at large are created with the common law and through statutes. Although the law of obligations has three components, contract, tort and restitution, the last only comes in to play in exceptional circumstances and so has not been covered in this chapter. Parliament's interventions through Acts of parliament are rare in what is essentially an area of private law (employment law and the rights of employees in particular are an exception).

The profession sets rules of conduct and competence that exceed the requirements of the common law. This is a good example of how professions set high standards that are commensurate with their status in society and reinforce the idea that the public can rely on the professions for their skill and competence.

There are certain basic principles that help to define a contract – but that these are not set in legal stone. These can be summarised as follows. There has to be an agreement; this is central to the idea of a contract. There has to be 'consideration'; this concept can be more readily understood as the effort required to perform a service – in our case design – in return for payment. And there has to be 'the intention to create legal relationships', the principle that makes it possible to enforce a contract.

Contracts are made to be performed and are not over until they are performed completely. A key aim of contract law is certainty and in the complex, fluid world of design services and construction contracts the only way to achieve this is to agree everything explicitly in writing. You should adopt this as a guiding principle.

The wider obligations to society are covered by the law of torts – negligence in particular. When we are negligent, we breach our duty of care to other members of society. Because these obligations are so wide-ranging a set of principles again help to define the level of responsibility. There must be a cause and an effect and the effect must result in physical damage or harm. However, because citizens rely on professionals, there are certain circumstances in which professionals giving negligent advice may be liable for economic loss as well as physical damage or harm. Again, Parliament has stepped in when the law is considered not to be working effectively, and this is particularly true of compensation for physical injury.

The common law of obligations has naturally developed around the disputes that people bring to a neutral forum in order to find a solution. Today, however, only a very small percentage of disputes end up in court. The rest are settled through negotiation or methods of alternative dispute resolution such as mediation.

FURTHER READING

Murdoch and Hughes Construction Contracts

Uff J 'Construction Law' (10th ed.) Sweet & Maxwell 2009

McKendrick E 'Contract Law' (9th ed.) Palgrave Macmillan 2011

Weir T 'An Introduction to Tort Law' Oxford University Press 2006

Martin J 'The English Legal System' Hodder 2010

Slapper G 'The English Legal System' (3rd ed.) Routledge 2012

Wevill J 'Law in Practice' RIBA Publishing 2012

7
PLANNING FOR THE FUTURE

'A map of the world that does not contain utopia is not worth looking at. For it is the one place that man is always landing.'[54] Oscar Wilde

INTRODUCTION

The central aim of this chapter is to introduce you to the system of regulation that has developed over the last century to control urban development. From a purely architectural point of view, this control can often seem like bureaucratic nonsense that gets in the way of free expression and realizing the client's vision. This view is blinkered, however: architecture has a powerful impact that cannot be considered in isolation from policy imperatives. An understanding of the history of planning control will open your eyes, hopefully, and temper your purist's instincts.

The built environment is now closely controlled by the legal constraints put in place by successive governments. Traditionally how land was developed and buildings designed were determined by the physical constraints of the site, climate and who owned the land. The common law traditionally supported the rights of landowners to make their own choices. Parliament was made up of landowners – who also had the right to vote. Not surprisingly, Parliament avoided any control over land use. Even if it wanted to, government and local government lacked the resources to exercise any control until the late nineteenth century.

This chapter focuses on the pressures that changed this over time and how Parliament 'stepped in' and eroded the common law rights in order to influence the way the physical environment is used, developed and more recently, preserved. As designers of the built environment you need to be aware of both the context for control and development as well as working with and responding to the physical environment itself in your design work.

Government policy towards urban development over this period has been driven by a number of factors. Traditionally called 'Town Planning', the first government interventions were driven by the need to improve the standard and layout of social housing in response to failures of existing housing provision in the fast-growing cities of the late nineteenth century. Later policies tried to improve the planning of towns to restrict urban sprawl and protect the countryside. Government then saw the potential of planning policy to shift the centres of economic activity in the country at large by encouraging development through controlling land use nationally.

The system that has emerged now balances government economic policy with the ability of local government to simultaneously encourage, control and restrict development. The trade-off for more political intervention has been to accommodate the rights of citizens through local democratic processes to influence how development strategies are defined and then to have a voice at local level. Architects have to work within this system but as experts designing within the urban environment we can proactively influence future national and local policy.

This chapter concentrates on the growth of town planning policy. The next chapter examines how governments have developed policies to protect our built heritage and landscape and sustainability.

THE BEGINNINGS: PLANNING POLICY 1890–1918

The main thrust of early planning legislation was the provision of affordable housing. The Housing of the Working Classes Act 1890 is not generally considered a piece of planning legislation however it is one of the first pieces of legislation that proposed a planned approach to the provision of 'working class housing'. It was a response to the acute housing conditions in city centres. One part of the Act gave specific powers to local authorities, if they wanted to adopt them, to provide what would now be termed social housing. Local government was ill-equipped to implement the policy, even if it wanted to. One exception was the London County Council (LCC) – see box below. The 1890 Act is important because the ideas developed by progressive local authorities such as the LCC influenced future planning legislation.

The London County Council was formed in 1889 with an elected body of councillors and an administration covering the area now classified as inner London. The LCC began by building a number of high-density social housing developments on sites it had acquired for the purpose. One of the earliest was the Millbank Estate, built on a former prison site next the Thames in central west London. Other buildings on the site include the Tate Gallery (example of nineteenth century philanthropy, in this instance by the sugar magnate Henry Tate) and the Millbank barracks – now an arts facility. It was one of the few examples of an available site of significant size in public ownership in central London. The LCC had the organisation and capability to make policy work. It had its own architects and also designed other building types such as fire stations and schools.

The Second Boer War (1899–1902) had surprising ramifications for planning policy. Although this conflict is overshadowed by the First World War nowadays, it was important at the time. The war in South Africa was not a great success in many ways. A Royal Commission[55] was set up under pressure from the establishment to task the government with investigating the reasons for failings in the military campaign. This was a major source of embarrassment but was also of great relevance to the country as the British Empire depended, in part, on Britain's military strength. One of the major problems for the Army had been recruitment – the Royal Army Medical Corps, who carried out medical examinations of recruits, found that forty percent of the men called up for service were physically unfit for military service. In the evidence given to the Commission their poor health was partly attributed to their living conditions and poor urban working class housing provision in particular. In an era before antibiotics, many medical 'cures' depended on a change of environmental conditions – 'fresh air' or a 'warm climate', for example. The power of improved social housing together with better sanitation and better working conditions to improve health was becoming recognized.

Government responded by passing the first Town and Country Planning Act (1909). The Act was a re-working of a piece of earlier private legislation created by the Garden City/Suburb movement to allow the development of the first Garden City at Letchworth and Hampstead Garden Suburb. Here was a case where political will aligned with theory – the Garden City ideals of Ebenezer Howard – to create policy. The architects for Letchworth and Hampstead Garden Suburb, Parker and Unwin, also published 'Town Planning in Practice' (1909) an influential practical guide and manifesto for their low-density, contextual approach to urban design. The 1909 Act was also one of a small number of relatively enlightened pieces of Edwardian Liberal legislation, including pensions, which lay the foundations for the future welfare state.

'The object of the Bill is to provide a domestic condition for the people in which their physical health, their morals, their character and their whole social condition can be improved by what we hope to secure in this Bill. The Bill aims in broad outline at, and hopes to secure, the home healthy, the house beautiful, the town pleasant, the city dignified and the suburb salubrious.'

John Burns, President of the Local Government Board[56]

The key purpose of the Act was to raise standards and it gave local authorities the powers to prepare town planning schemes to control the development of new housing. The consultative and approval process was cumbersome, requiring approval of schemes by the Local Government Board.

The impact of the Act was cut short by the First World War but in the period between 1910 and 1914 the LCC, in particular, built three Garden Suburbs on the outskirts of London at Norbury in South London, Old Oak in West London and White Hart Lane in Tottenham in North London. These schemes were restricted to housing only as the LCC did not have the powers to create facilities to serve the community such as shops or places of employment. The private sector was not similarly restricted and the new communities at Letchworth Garden City and Hampstead Garden Suburb continued to be developed with a range of facilities and different uses such as shops and libraries.

During the First World War the architects employed on designing Letchworth and Hampstead Garden Suburb were re-employed on government schemes. The most significant contribution in town planning terms was the relatively large-scale development of housing for munitions workers at Well Hall Woolwich and Gretna in Scotland which saw the application of Garden City/Suburb principles to social housing. Raymond Unwin (of Parker and Unwin) worked on Gretna, built in a restrained striped Georgian style whereas Well Hall was built in an eclectic Arts and Crafts style using available materials. Unwin also helped to draft the policy for standards of post-war social housing recommended in the Tudor Walters Report of 1918, a government report that gave guidance on the planning and design of housing schemes and minimum standards for housing density.

PLANNING POLICY 1919–1940: FROM HOUSING FOR WAR HEROES TO TACKLING URBAN SPRAWL

The key objective of planning policy after the First World War was the provision of affordable housing. This was a political objective: a response to public sentiment and the cornerstone of Lloyd George's successful election campaign promising 'Homes for Heroes' which helped him to office after the end of the First World War.

During the 1920s, urban development followed three different and contrasting paths:

1. Carefully planned new local authority housing developments built to set standards and on Garden City principles.

2. The Garden City – based on zoning and social provision which would hopefully be the model for future new communities.

3. Uncontrolled low density private sector development on the farmland around towns and cities.

The first path saw social housing as a priority. The 1919 Housing and Town Planning Act (the 'Addison Act') established the principle of state subsidy for new housing. All local authorities above a certain population had to prepare schemes but the emphasis was on housing rather than urban design. The new standards also meant that social housing could only be effectively provided on the periphery of towns following a low-density Garden Suburb model to a low density of 12 houses per acre as recommended in the Tudor Walters report. This was relatively cost-effective: growing existing communities located in established commercial centres. Unwin became their chief architect of the Ministry of Health which took responsibility for social housing and was joined by another Hampstead Garden Suburb architect, Michael Bunney.

The second path saw the solution in the Garden City model of new, self-contained communities of approximately 30,000 inhabitants. Ebenezer Howard set about organizing and planning the next self-contained community. Land was obtained and Welwyn Garden City was planned to an original masterplan by another Suburb architect, Courtney Crickmer. Welwyn was to be funded using the private sector, philanthropic model with zoned commercial, industrial and residential sectors linked by public spaces and close to a major railway artery. Clearly this model was much more expensive than the solution employed by the Ministry of Health as it required entirely new infrastructure. It also required greater co-ordination of different private sector enterprises from industry to locate new factories there to generate employment to public sector administration of services. As a result progress was painfully slow.

In contrast, private sector suburbanization, the third path, accelerated – driven by rapid improvements in public transport and the opportunity to satisfy the demand for new middle-class housing. This is exemplified in 'Metroland' the name given to the almost unrestrained suburban development made possible by the growth of the Metropolitan Railway into rural Middlesex and Buckinghamshire.

Attempts were made to control the rapid expansion of suburbanization. The Town and Country Planning Act 1932 extended planning powers and the Ribbon Development Act 1935 attempted to control development along major roads. However, the administration was cumbersome. Under the 1932 Act local authority schemes took three years to prepare and Parliamentary approval was also required. The schemes zoned land for particular purposes such as industrial and residential. Most schemes did little more than confirm existing uses, as anything that imposed different uses would have required local authorities to pay compensation to land owners. The administration by local authorities was also weak: generally they did not have the capability or competence to implement coherent schemes. The problem got worse during the housing boom of the 1930s: 2.7 million houses were built in England and Wales between 1930 and 1940.[57] Urbanization was not evenly spread across the country. The population of Greater London rose by two million and the population in the North-East and Wales fell. Unemployment levels were also significantly higher in the North-East than in the South.

So, new issues had arisen: the shift in population away from the regions, unemployment 'blackspots', the growth of London, unchecked suburbanization and the loss of countryside (particularly high-grade agricultural land), to housing development. As the 1930s proceeded central government recognized that existing legislation and initiatives designed to stimulate employment were not effective.

Up till now, the development of statutory town planning was essentially responsive to major events but underpinned by certain ideas and trends:

- Improvements to living conditions and housing in particular, can be a powerful agent for social good.

- Limited government intervention can influence and drive change.

- The way to execute change is to delegate planning decisions to local authority level (local authorities were tasked to produce schemes for housing provision).

- Central government retains control by sanctioning approval of these schemes.

- Government subsidy can act as an incentive and direct building activity towards the areas of need.

- The Garden City approach to urban design and town planning promoted by Howard is sidelined in favour of treating each problem separately.

This emerging practice had a number of weaknesses.

- **Structural weakness.** Delegating the planning to local authorities and retaining central government control set up a tension between the two areas of government where, at its worst, perceived local need could be over-ridden by central government objectives. The process was also inherently slow and potentially inefficient with resources committed to schemes that would be rejected. The lack of central co-ordination made it difficult to achieve an even national impact.

- **Insecure subsidies and incentives.** These could be turned on and off like a tap. Therefore long-term plans could be changed to suit wider economic (and social) policies

- **Lack of resources**. Local authorities lacked the expertise and powers to make effective local plans.

- **The policy of zoning.** Zoning and locating different uses (industrial, residential) proved an ineffective way of controlling development or stimulating employment.

The task of co-ordinating government intervention to stimulate employment and control the even distribution of production proved a very difficult task. In response the government appointed another study: the Royal Commission on the Distribution of the Industrial Population. The Second World War broke out shortly afterwards.

PLANNING A NEW UTOPIA: 1940–1970

A pioneering report, the Barlow Report was published in January 1940, shortly after the outbreak of the Second World War. It is important because it identified the deficiencies of the pre-war system and made a series of proposals that set out the principles for post-war regional planning policy and set the template for government planning up to the present day.[58] It proposed a central authority to co-ordinate the dispersal of industry and population from congested urban areas using a variety of solutions where appropriate: garden cities and satellite towns, the expansion of smaller regional towns and garden suburbs and the diversification of industry within each region.

Further reports followed, demonstrating optimism about the future:

- The Scott Report on Rural Land Use (1942), which argued for the conservation of agricultural land.

- The Abercrombie Greater London Plan (1944), which introduced physical development plans for London (with the intention of moving 1.5m people out of London to new settlements).

- The Reith Report on New Towns (1946), which argued that the physical planning of new towns could create a new community with appropriate amenities.[59]

These reports embodied common principles which can be summarized as:

- A national policy that would use the planning system and especially zoning and land use as a lever of economic policy.

- The planned dispersal of industry and population; the containment of existing urban centres, especially London.

- The preservation of agricultural land as a priceless asset based on a fairly crude distinction between the 'urban' and 'rural' environment.

- Planned expansion of regional centres and 'new town' communities.

The destruction of major cities, such as Coventry, in the *'Baedeker Raids'*[60] during the war gave urban designers and architects the opportunity to plan a better future. They used contemporary urban design principles such as zoning and urban containment, combined with a British interpretation of European Modernist thinking. The raids created a 'tabula rasa' in many cases, or gave city councils the opportunity to create one. For example, by 1942 Plymouth, which had been badly damaged, had a new plan for the City. In Hull, Abercrombie and Lutyens (1945) argued for a new urban structure on which to re-build the traditions of the great port. In Exeter, (1947) a new plan was proposed for the City.

The 1947 Town and Country Planning Act effectively nationalized almost all development and land use – but the mechanisms for making government policy happen were fragmented. One department was responsible for development and land use while another was responsible for the control and incentivisation of new industrial projects. New Towns were developed by separate individual corporations. National Parks and Areas of Outstanding Natural Beauty were the responsibility of a separate body. Broadly, and subject to the rationalizations discussed briefly below, the same systems and fragmented structures remain in place today. (Agriculture, and development linked to farming remained outside the planning system.)

During the 1950s the urban design schemes that were proposed as the physical embodiment of regeneration policy shared many strategic and stylistic features. Strategically, urban design offered the opportunity to correct the perceived errors of the past and to create a better future. Their architectural and urban design represented a watered-down version of the very British response to European Modernism – mixing relatively new materials, construction methods and spatial design with an eclectic mix of design features that often tried to reinterpret the past. Cities such as Canterbury and Plymouth and parts of bombed out London took their cue from the new-found optimism of the new towns with sometimes quirky references to the stylistic features of the 1951 Festival of Britain.

Green Belts

Of all planning policies, Green Belts are the most widely-recognised and most popular. They were first proposed in the pre-war period in response to unrestrained suburban development and the first areas were designated around London in 1938. The 1947 Act also made a provision for them and in 1955 the policy was extended to all major urban areas. The objectives were to check the growth of built-up areas and prevent towns merging into one another. The1955 Government Circular is still in force today and forms part of the policy set out by the government.[61] Currently 13% of England, 2% of Scotland and 16% of Northern Ireland are designated Green Belt land. Policies are in place to establish Green Belts in Wales.[62]

'Traffic in Towns' 1963

'Most of our people have never had it so good.'

Harold Macmillan, Prime Minister. 20*th* July 1957[63]

Along with the post-war 'baby boom' car ownership increased as did traffic congestion. In 1963 the 'Buchanan Report' 'Traffic in Towns' confirmed the potential damage caused by the car and suggested ways to deal with it.[64] Ahead of its time, the report discussed the environmental impact of the car and recommended restricting car usage, including segregating pedestrians from traffic and building ring roads. Buchanan gave a number of speculative examples illustrating how the principles of segregation could work. For example, one option for the congestion in Oxford Street, London was to build roads in shallow cuttings, underground car parks and raised access to shops with high-level pedestrianised walkways. Buchanan even proposed a similar scheme for Bath.

In the dash to accommodate the motorist by a process of 'predict and provide', the balance between providing for the car and mitigating its environmental impact was lost. The legacy includes: pedestrian precincts, urban clearways, flyovers, multi-storey car parks and road planning becoming a major function of urban design. It shaped (or destroyed, depending on your position) most major city centres in the UK.

The link between traffic improvements and wholesale development was not lost on developers who were able to propose large-scale reconstruction linked to major traffic improvements.[65]

Unfortunately the approach to traffic improvements was also enthusiastically adopted in large and small towns. For example, in Chesham in semi-rural Buckinghamshire historic buildings were bulldozed to make way for new traffic schemes and parking. This destruction of the urban grain and fabric led belatedly to an evaluation of the merits of the past as well as a questioning of the promises for the future that will be examined in the next chapter.

CONSOLIDATION – THE SYSTEM MATURES 1970–2012

The period since 1970 has seen a significant number of changes to the statutory planning system. The growth of heritage and conservation legislation that also developed over this period is discussed separately in the following chapter. Regional and local government have been through a number of re-organisations. The political use and interpretation of the planning system has lurched from an instrument of interventionist control to a brake on 'free enterprise' and back again a number of times depending on the government at the time. However over this period the planning system has consolidated around the 'plan-based system', the principle of developing local statutory plans which are reviewed through consultation in a local democratic forum. Local planning policy was also influenced by wider central government directives (such as regional housing targets) and departmental planning guidance and statements (Planning Policy Statements, PPSs) which set out a national policy for particular areas of interest. In 2010 there were twenty-two of these in total ranging from Sustainable Development and Climate Change (PPS1) to Green Belts (PPG2) and Noise (PPG24).[66]

Successive governments have tried to streamline the planning system but until recently seemed reluctant to remove controls. With each change of government another layer of complexity is added.

Added Complexity

For example in the 1980s the Thatcher government published 'Lifting the Burden' – which speaks for itself. At a practical level the government introduced 'Use Classes' as a way of simplifying the system. Changing use was deemed to be 'development'. The new system said that you could change use within a category without applying for planning permission. However, these might be open to interpretation, or a new use class could develop – such as 'live-work' which did not fit into an existing category. Zones free from development control were also set up with the objective of encouraging enterprise and development.

The late 1990s through to 2008 saw a sustained period of economic growth. As time went on the perceived drawbacks of the planning system became a *cause celebre'*. A system designed to encourage but control growth and act as a lever of economic policy was now seen as a system that restricted growth. Central government used planning policy to target growth and housing in particular and, following the effective withdrawal of government from the provision of social housing, used regional spatial planning policies as a way of facilitating housing growth – linked to the private sector. The policy consistently failed to deliver, leading to greater housing shortages and spiralling house prices fuelled by readily available borrowing.

'ALL CHANGE'

Following the economic crash of 2008 the planning system was again blamed for holding back economic recovery. The construction industry represents approximately 6–7% of what the country produces ('GDP': Gross Domestic Product). Therefore anything that allows it to grow will, in theory, encourage economic growth too. The election of the Coalition government led to a new approach that dismantled the regional planning policies and centralized housing targets of the previous Labour administration. This was replaced by 'Localism' with the aim of strengthening local democracy in planning decision-making.

In 2012 a new policy was published: the 'National Planning Policy Framework' (NPPF).

The key elements of the NPPF are:

■ enshrining the local plan – produced by local people – as the keystone of the planning system,

■ making planning much simpler and more accessible – reducing over 1,300 pages of often impenetrable jargon in 44 separate documents into a clear, readable guide of 50 pages,

■ establishing a powerful presumption in favour of sustainable development that underpins all local plans and decisions,

■ guaranteeing robust protections for our natural and historic environment, including the Green Belt, National Parks, Areas of Outstanding Natural Beauty and Sites of Special Scientific Interest,

■ encouraging the use of brownfield land in a way determined locally,

■ in addition, the new Framework strengthens the requirement for new development to be of good design; supports local councils who wish to bring into being a new generation of garden cities; allows communities to specify where renewable energy such as wind farms should, and should not, be located.[67]

The system of planning policy statements (PPSs) was dismantled overnight, without questioning whether they had been of any use in helping to guide local authorities and the public through the complexities of sixty years of planning legislation. Its main aim was to make the planning system simpler, locally-based and more accessible, to protect the environment and to promote sustainable growth.[68] The major and most controversial change was the new starting point: all planning policy – local as well as national would be led by the principle of 'sustainable development' unless there are national and local policies in place that state otherwise. The Green Belt and historic buildings and highly-valued natural environments will still be protected but the principle of 'sustainable development' rules elsewhere. It is interesting to see a new reference to garden cities. It remains to be seen if it will provide a stimulus to the national economy whilst preserving local values.

Although PPSs have generally been cancelled, some remain. These include guidance on 'Eco-towns', Waste Management, Flood Risk and Coastal Change. It will be interesting to see how the new system evolves and matures in the future and whether the parallel system of heritage controls discussed in the next chapter remains intact.

CONCLUSION

This overview of the growth of town planning has brought together a number of different strands and is an example of how national government has successively tried to influence land use and urban development and use it as an instrument of both economic and social policy. Even with the new policies of 'Localism' and 'Sustainable Development' in place the urban and rural environment will still be subject to a regulatory framework which will affect both large-scale and small-scale decision-making. Architects and designers will need to be fully aware of current trends and legislation in order to navigate their way through the system – as well as to shape future policy.

FURTHER READING

Cullingworth B & Nadin V *'Town & Country Planning in the UK'* (14th ed.) Routledge 2006

Hall P & Tewdwr-Jones M *Urban & Regional Planning* (5th ed.) Routledge 2011

Rydin Y *'The Purpose of Planning'* Policy Press 2011

8
SAVING THE PAST
– AND THE PLANET

'Nostalgia is not what it used to be.' anon

INTRODUCTION

Heritage

This chapter looks at how changing attitudes to our spaces and places has been reflected in the regulatory framework that controls our heritage and the environment. This starts with a concern for ancient monuments to buildings of historic and architectural interest to the statutory protection of the countryside and urban areas. More recently the government has turned its interest to the wider environment and its response to climate change and sustainability. This chapter aims to explain this complex and rapidly changing legislative environment under three broad headings: heritage, environmental protection and sustainable development.

Definitions

Key words in this discussion include preservation, conservation, heritage and sustainable development. There is often confusion about their meaning – in part due to the use or misuse of key phrases by politicians and the media to play on the more emotive aspects of the environment.

◼ 'Preservation' is the oldest concept and implies retention without significant change and protection in the original or close to original state.

◼ 'Conservation' is a twentieth century concept which suggests change that respects and retains key elements of recognised value.

◼ 'Heritage', in this context, suggests: 'a process of evaluation, selection and interpretation – perhaps even exploitation – of the past'.[69]

◼ The meaning of 'Sustainable Development' has been diluted by over-use but may be seen as the balance between economic development and environmental conservation.

THE ORIGINS OF 'HERITAGE': CHANGING ATTITUDES

In contrast to the international growth of urban design theory and the socio-political as well as spatial theories that underpin urban design it is difficult to identify a body of theory that drives the heritage movement. There is a perception that urban design theory is international, whereas attitudes to heritage appear to grow from a sense of national values and history. For example, when France began to review its approach to conserving the urban environment in the early 1960s it turned to a socialist philosopher, Malraux, to create a national conservation policy that recognised the 'atmosphere' of the historic urban environment as well as formulating an approach to conserving the historic fabric. The origins of the contemporary English attitude to heritage can be traced back to the intellectual and artistic reaction to industrialisation in the late nineteenth century and also the more sophisticated critique by authors such as Thomas Hardy, an architect before becoming a celebrated author, of the 'restoration' of medieval churches which led to the wholesale destruction of the historic fabric and a re-invention of gothic church architecture as a sanitised version of the past which aligned with the High Victorian view of institutionalised religion.

The statutory response to heritage over time can be interpreted as a response to changing awareness and approaches to buildings as artefacts, the historic urban grain of cities and towns and more complex attitudes about the countryside. The growth of heritage legislation is almost completely independent of the growth of town planning legislation. You will know from the discussion in the last chapter that town planning legislation is broadly driven by socio-political factors such as the need for housing and economic growth which led to a plan-based system executed at a local level. The legislation is intended to promote growth and development within a framework of democratic controls. Heritage legislation takes protection, preservation and conservation as its starting points. The system therefore runs in parallel and often in conflict with development control. Environmental protection legislation, in turn runs in a further parallel stream with the emphasis on compliance with control partly driven by 'top-down' European legislation intended to mitigate the effects of global warming.

FROM PRESERVATION TO CONSERVATION

The late nineteenth century saw limited efforts to protect our heritage as represented by ancient monuments such as Stonehenge in rural Wiltshire built in approximately 3100 BC and Hadrian's Wall in Northumberland built by the Romans. The first efforts to preserve our historic monuments were driven by a widening interest in history and archaeology coupled with a reaction to the crude 'restoration' work carried out by Victorian architects, a neglected degradation of the countryside caused by a series of agricultural depressions, rural depopulation and the rapid expansion of towns and cities that was swallowing up the historic landscape.

The voluntary sector was then, as now, critical to supporting the process of preservation and creating a cultural attitude to the historic fabric. The Society for the Preservation of Ancient Buildings (SPAB) was created by William Morris and other prominent members of the Pre-Raphaelites brotherhood in 1877 concerned that over-zealous architects were scraping away the historic fabric of buildings in the name of 'restoration'. The SPAB, as early as 1896 co-hosted a conference with the London County Council on the preservation of ancient buildings in London which raised awareness of the threat to the historic fabric and resolved to create a register of threatened buildings as a first step to protection and preservation.[70] During the 1930s it continued to raise awareness of buildings and structures under threat – from bridges threatened by road improvement schemes to water and windmills. At the same time it also formed a Georgian Group to concentrate on this important sector of British architecture and it continues to add to the body of conservation scholarship and education.

'The National Trust for Places of Historic Interest or Natural Beauty' was founded in 1884 as a private association. The Trust soon adopted the SPAB's approach to conservative repair rather than 'restoration'. In 1907 an Act of Parliament (The National Trust Act) was introduced to change the structure of the organisation:

'The National Trust shall be established for the purposes of promoting the permanent preservation for the benefit of the nation of lands and tenements (including buildings) of beauty or historic interest and as regards lands for the preservation (so far as practicable) of their natural aspect features and animal and plant life.'[71]

Although we now see these ancient monuments as part of our national heritage and therefore as *public* monuments they were mostly on private land and effectively privately owned. Inevitably landowners objected that their common law rights were being eroded. The Ancient Monuments Act 1882, a watered down version of efforts to introduce legislation in the 1870s, acknowledged that it was in the national interest to preserve ancient monuments but any actual preservation was more to do with the cooperation and goodwill of enlightened landowners than any measures of statutory protection.[72] In short, protection and preservation could only be reliably achieved by either bringing the land into public ownership or imposing controls. Neither was practical or politically acceptable. A public record of historic monuments, though, was more achievable. This idea of systematically recording and studying monuments mirrored the then-current museum practice: a place to catalogue, preserve and display – an extension of the eighteenth century idea of 'the cabinet of curiosities' that eventually led to the foundation of institutions such as the British Museum. However the scale was significantly different and cataloguing these objects, the origins of which were sometimes misunderstood, was itself a major task. A major step forward was the creation in 1908 of the Royal Commission on the Historic Monuments of England, Scotland and Wales whose purpose (in the case of England) was:

'to make an inventory of the Ancient and Historical Monuments and constructions connected with or illustrative of the contemporary culture, civilisation and conditions of life of the people of England from the earliest time to the year 1700 and to specify those that seem most worthy of preservation.'[73]

You will see that the emphasis is on recording and *then* possible preservation. The cut-off date was 1700 – reflecting attitudes of the time which have been mirrored in more recent years – that objects become more valuable with time. In 1913 powers were given to local authorities (or the Commissioners) to purchase an ancient monument allowing them to become 'guardians' – thereby preventing destruction of a monument – while leaving the 'ownership' in private hands. This idea of monuments as 'things' of historic value in their own right, especially if catalogued, tends to exclude the context of these objects, the setting and place, the gaps in-between, that is so important to an understanding of the urban grain of towns, cities and landscapes and hindered a more complex holistic approach to conservation.

In the 1940s it was decided to establish a national survey of historic buildings – a huge task. This 'listing' of buildings 'of special or historic interest' was held centrally and was separate from the provisions for ancient monuments. The list was given extra teeth when listed buildings were given protection in 1968 by the introduction of a system where Listed Building Consent was required for the demolition or alteration of a listed building. The system of permissions given by the local authority is separate from, and runs in parallel to, the system of development control. This protection of historic buildings was in part prompted by the publicity given to the destruction of significant buildings such as the classical-style Euston Arch outside Euston Station in London to make way for an American-inspired 'plaza' project of office towers and a new station building. The 'Euston Arch' still holds a mythological value for conservationists – a sort of holy grail – with conspiracy theories fuelled by the hope that the parts will one day be discovered and that it will be reconstructed.

What makes a Listed Building?[74]

The criteria for listing reflects Society's changing attitudes towards protection and conservation. As English Heritage, the government body responsible for listing, states:

'Listing helps us acknowledge and understand our shared history. It marks and celebrates a building's special architectural and historic interest, and also brings it under the consideration of the planning system so that some thought will be taken about its future.'

In principle the older the property the more likely it is to be listed. These are the broad criteria used by English Heritage:

■ Age and rarity: most buildings built before 1700 which survive in anything like their original condition are listed, as are most of those built between 1700 and 1840.

■ The criteria become tighter with time, so that buildings built within the last 30 years have to be exceptionally important to be listed, and under threat too. A building has to be over 10 years old to be eligible for listing.

■ Architectural interest: buildings which are nationally important for the interest of their architectural design, decoration and craftsmanship; also important examples of particular building types and techniques.

■ Historic interest: this includes buildings which illustrate important aspects of the nation's social, economic, cultural or military history.

■ Close historical association with nationally important people or events.

■ Group value, especially where buildings are part of an important architectural or historic group or are a fine example of planning (such as squares, terraces and model villages).

Listing is in three (descending) categories Grade 1, Grade 2* and Grade 2

Anything built since 1945 has to be exceptional to be listed. This includes buildings such as the Norman Foster's Willis Faber Dumas building in Ipswich (1972–1975) and the Severn Bridge. In England in 2012 there were 374,081 entries (some of which relate to multiple buildings such as terraces of houses). Only 424 or 0.2% of the total were built after 1945.

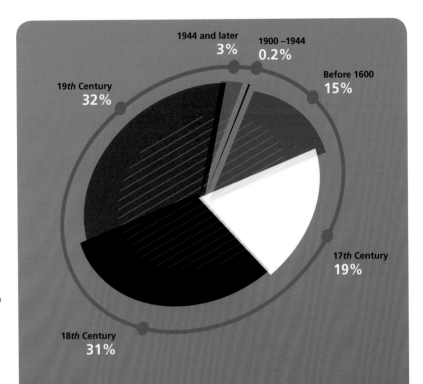

1944 and later
3%

1900 –1944
0.2%

Before 1600
15%

19th Century
32%

17th Century
19%

18th Century
31%

FIGURE 8.1

Graph detailing the age range of listed buildings in the UK[75]

Once listed, buildings are protected from unauthorised demolition or alteration. To carry out works without permission is a criminal offence – as a select group of owners, their architects and builders know to their cost. Depending on the grade alterations to a listed building are controlled through the local planning system using a system that runs in parallel, and often in conflict with, the normal planning system of development control. Therefore owners wanting to alter or extend a listed building will require planning permission and Listed Building Consent. There is always the possibility that they may get one and not the other.

To complicate matters further local authorities also prepare their own lists of 'locally listed' buildings of historic or architectural interest which can be integrated into local planning policy. The intention is that the local population has a voice in the designation.

In addition to listed buildings English Heritage also has:

- 19,717 scheduled ancient monuments,
- 1,601 registered historic parks and gardens,
- 43 registered historic battlefields,
- 46 designated wrecks,
- 17 World Heritage Sites.

FROM OBJECT TO PLACE TO SPACE: CONSERVING THE URBAN GRAIN

Listing perpetuated the nineteenth century idea of cataloguing and ordering objects. By its nature the process is highly selective and descriptive excluding urban design theory and practice and the idea that context can have its own worth. Under the listing system an urban area can only have historic value as a result of the grouping of individual buildings. This could be problematic when the listing system set an arbitrary age barrier for listing or ignored the contextual importance of buildings that did not fit the listing criteria.

There also needed to be a change in attitude by both the government and the public. As seen in the last chapter, the 1950s and early 1960s saw a drive to improve the housing stock which led to the destruction of 'sub-standard' housing under wholesale slum clearance policies and its replacement with new social housing. In the dash for growth and improvements in living standards subsidies were introduced which favoured high rise blocks of flats surrounded by green spaces once occupied by terraced housing. In the private sector the conservative attitudes of lenders favoured new-build properties and it was difficult to get loans for the refurbishment and improvement of the old housing stock.

The effects of 'Traffic in towns',[76] published in 1963, and which gave policy blueprints for how to deal with traffic congestion were also felt. The historic fabric was systematically destroyed in many towns and cities to make way for traffic improvement schemes. This re-planning also led to development opportunities for new shopping centres and commercial offices – often encouraged by local politicians and their planning departments and architects. Traffic schemes also had the effect of cutting through communities and severing inefficient but vital circulation routes that had grown up over time but now failed to fit in with a rational approach to transport and plan-led zoning of uses.

Gradually the public learnt to live with the policy changes. A new approach to area-based conservation was required and the government was heavily influenced by the system introduced in France by Malraux, the Minister of Culture (a term not adopted by UK government until the 1990s) in 1962. As already seen, the system designated historic areas for co-ordinated preservation and conservation – recognising the 'atmosphere' of historic areas as well as individual buildings of architectural merit. In the late 1960s local authorities in England and Wales were given a statutory duty to designate conservation areas – groups or streets or small areas in old towns. Within these areas buildings were effectively protected and permission was required for the demolition and alteration of buildings.

It was originally envisaged that only a few pre-Victorian areas would be protected. Over time the number of Conservation Areas, as they are known, has grown. This has been almost entirely local authority officer-led and the designation of areas is neither democratic nor particularly rigorous. The control of development in town centres has been strongly criticised as contributing to the ossification of local economies. At the same time developers, restrained by these controls, have developed shopping centres on the edges of towns where there are fewer restrictions and abundant free parking. In 2012 there were 9,080 conservation areas in England.

There are many different types including:

- the centres of our historic towns and cities,

- fishing and mining villages,

- eighteenth and nineteenth-century suburbs,

- model housing estates,

- country houses set in their historic parks,

- historic transport links and their environs, such as stretches of canal[77].

When a local authority creates a Conservation Area, normal development rights are suspended and Conservation Area Consent is required from the local authority. This may include relatively minor alterations to the external appearance such as changing windows or installing a satellite dish – even the colour of your front door. Trees are also protected with consent required for pruning as well as removal.

Case Study: Covent Garden

In the late 1960s a new plan to move the historic flower and fruit market in Covent Garden to Nine Elms in South London meant that the existing site would become vacant. The new plan for the area brought together the doctrines for segregated pedestrians and traffic improvements set out in Buchanan's Traffic in Towns together with new commercial developments. The ensuing battle formed a watershed in changing attitudes to urban planning and the value of conservation areas.

THE BATTLE FOR COVENT GARDEN

If the original plans for Covent Garden had been implemented, two thirds of the area would have been demolished and replaced with high rise buildings, 'pedways' [segregated pedestrian walkways] and an underground ring-road.

This now inconceivable scenario was only thwarted after a historic battle by residents and businesses led to over 200 buildings being 'listed' by the late Geoffrey Rippon, Secretary of State for the Environment. These events helped turn the tide of post-war planning in the UK away from wholesale demolition and decanting residents, to a more sensitive approach to our historic city centres, and to London's long standing residential villages such as Covent Garden, Soho and Bloomsbury.[78]

The battle also attracted different factions who united in their efforts to defeat the plans. This threw up some unlikely heroes as exemplified by this obituary of the architect and left-wing activist, architect and teacher, Brian Anson, written by the architect and urban designer, Richard Rogers.

'Brian will be remembered for his role in the fight for Covent Garden in the late 1960s. He lost his job at the Greater London council as a result of siding with local residents against the council's plans to tear down the historic buildings and install a monstrous, car-dominated redevelopment. It was an epic battle between developer and citizen, a pattern then occurring across Europe, but Brian and a group of friends prevailed against the odds. Had they lost, London today would be a less humane and beautiful city.'[79]

The publicity and popularity created by the Battle for Covent Garden formed the template for public participation in future planning battles and helped to heighten the awareness of the vulnerability of the historic fabric and urban grain.

THE GROWTH OF HERITAGE

Heritage has been taken up by society in a way that the more academic, scholarly approach to cataloguing, preserving and conservation has not. Conservation and heritage although often confused as the same thing are not. Perhaps as a result of its universality, heritage is ill-defined but nevertheless remains as a powerful concept – if only because it appears to unite many different ideas about the past and national identity.

'Heritage is neither history nor place: it is a process of selection and presentation of aspects of both, for popular consumption. Heritage is history processed through mythology, ideology, nationalism, local pride, romantic ideas, or just plain marketing, into a commodity.'[80]

Heritage is a useful concept because it can accommodate some of the ambivalence we have about planning for growth and conserving the past. Although the business community and many politicians complain that the planning system is holding back economic activity the heritage movement can point to the power of heritage schemes to rejuvenate areas such as the Grade 1 listed Albert Dock in Liverpool and Covent Garden in London and fuel the local economy. Nationally, tourist authorities promote Britain's past – rather than its weather to international tourists. The media interest in the past – where building projects are represented as part-narrative of the triumph of the underdog over authority and part-detective story has raised the awareness of buildings at risk and reflect changes in values and our attitudes to the past. This increase in interest has also been affected by increases in living standards, leisure time and possibly the age profile of the population (who wield the 'Grey pound'). Although the government supports heritage through limited grants and the Heritage Lottery Fund the majority of heritage projects are driven by volunteers working under the aegis of organisations such as the National Trust.

Heritage is big business

The concept of heritage is hugely popular in the UK. Organisations to protect it have impressive numbers of members. In 2012 there were over 3.7 million members of the National Trust. The Trust whose strapline is 'forever for everyone' looks after over 350 historic properties, gardens and ancient monuments that are open to the public. In 2011 17 million people paid to visit these and an estimated 50 million visited the open spaces such as coastline forests and fens that it also looks after. The National Trust is an independent charity and relies upon donations, entry and membership fees for its income. The Trust in turn relies upon an army of 61,000 volunteers, a team of specialist staff and consultants including architects.[81]

Not to be outdone, English Heritage – a part of the government Department for Culture, Media and Sport – has 750,000 paid-up members and 400 properties open to the public which were visited by 11 million people in 2011.[82] Heritage is big business, so much so that English Heritage has teamed up with the music impresario Andrew Lloyd Webber – of Cats and Phantom of the Opera fame – to promote 'Heritage Angels' to recognise and celebrate the volunteers working to protect buildings at risk of neglect and demolition.

Behind the scenes thousands of voluntary organisations from the Society for the Protection of Birds to the Chiltern Society – set up to protect the Chiltern Hills north of London from development – work to protect aspects of our environment whilst adhoc pressure groups form and disappear to stop development ranging from London Heathrow's third runway (successful – for the time being) to trying to stop major infrastructure projects such as HS2 – the high-speed rail link connecting London to Birmingham.

Heritage is also international. The Unesco World Heritage Convention lists over 900 sites that 189 UN members are committed to protecting. These include cities as diverse as the old city of Sana'a in the Yemen to the Matobo Hills in Zimbabwe.[83] The list includes 28 sites in the UK from the Giants Causeway to the Tower of London. Sites must be of 'universal value' and meet one out of ten diverse criteria from 'representing a masterpiece of human creative genius' to 'natural habitats for insitu conservation of biological diversity.'[84]

The business of heritage is paradoxical and highly subjective. The historic centre of Warsaw, the capital of Poland, is a case in point.

'During the Warsaw Uprising in August 1944, more than 85% of Warsaw's historic centre was destroyed by Nazi troops. After the war, a… reconstruction campaign by its citizens resulted in today's meticulous restoration of the Old Town, with its churches, palaces and market-place. It is an outstanding example of a near-total reconstruction of a span of history covering the 13th to the 20*th* Century. ….This example illustrates the effectiveness of conservation activities in the second half of the 20*th* Century, which permitted the integral reconstruction of the complex urban ensemble.'[85]

In other words, the reconstruction of pre-1939 Warsaw – a selective process that required a reinterpretation of the past – has been recognised for its own intrinsic value. Of course, this raises the question: 'At what point do you 'freeze' the reconstruction?' Choices had to be made. Not only has the past been recreated but it has also been enhanced and sanitised. In Warsaw it is an artifice of selection, preservation, conservation and interpretation (based on research and careful reconstruction) – to give the *appearance* of a historic city centre rather than the authenticity of the preserved historic fabric that makes it an international heritage site. As such, UNESCO sees it as a model for late twentieth Century conservation practice. In the twenty-first Century, replica places and the skill which with they are (re)constructed may become as important as the original artefacts or urban grain.

ENVIRONMENTAL PROTECTION

The objective of protecting the natural environment has run hand-in-hand with the interest in urban planning. It shares the same origins – a concern for living and working conditions in Britain's towns and cities in the late nineteenth century. However legislation to protect the environment has not followed the same course as the plan-based system of development control or the protection of the historic urban fabric. (The Victorian concerns for public health are discussed in the next chapter and follow a different strand of protection and control.) At the heart of contemporary concerns are the waste and pollution created by development. Pollution initially concerned airborne and waterborne waste but now includes noise, light and the general degradation of the natural environment caused by development activity and its effects such as road and air traffic.

The plan-based system of development control has been relatively slow to recognise the consequences of development and planned urban growth. As seen in the last chapter Green Belts were seen as a way of protecting the countryside augmented by the creation of Areas of Outstanding Natural Beauty (AONB), Sites of Special Scientific Interest (SSSI) and the designation of National Parks – all of which serve to restrict development and preserve the natural landscape.

Although industrial pollution had been a well-established phenomenon since the beginning of the Industrial Revolution it took the 'Great Smog' in London in 1952 to kick-start efforts to control airborne pollution. Although the smog only lasted five days – and London was used to 'pea soupers' as these episodes of dense pollution were called, medical reports indicated that 4,000 people had died prematurely and 100,000 fell ill as a result of the pollution. The first Clean Air Act was introduced in 1956.

In the dash for post-war economic growth there was little concern for industrial waste and pollution or particular recognition of the link between the two. Pollution was categorised as an industrial problem and little attention was given to the discharge of sewage into waterways and coastal waters by the publically controlled utilities or the effects of pesticides and chemical fertilisers used in farming. The publication of 'Silent Spring' – the title referring to a countryside without birdsong, all killed by pesticides, in the USA in 1962 is seen as a watershed moment in environmental awareness and led to the banning of one of the most toxic pesticides DDT by the USA in 1972. As the science improved and the effects of pollution became more widely understood so public awareness also increased.

Measures to reduce pollution, such as the Control of Pollution Act 1975, generally developed independently of the planning system over this period. The catalyst for a more holistic approach was the European Directive (1985)[86] which required Environmental Impact Assessments (EIAs) for a wide range of public and private sector infrastructure projects such as roads and railways. This was further amended in 1997 to take account of the UN ECE[87] Espoo Convention of 1991 (in force from 1997) which required states to take a cross-border approach to environmental impact. In 2003 a new Directive included a requirement for public participation in decision-making and access to justice on environmental matters.[88]

The 1980s and 1990s saw the emergence of attempts to integrate different environment policies – kick-started by the 1982 UN Conference Environment and Development (the 'Rio conference' because it was held in Rio de Janeiro, Brazil). In 1990 the government white paper 'This Common Inheritance' linked environmental protection with the planning system and proposed that local authorities should consider all effects, including pollution, when granting planning permission.[89] The Environment Agency was created in 1995 to bring a science-based, integrated approach to pollution control covering land, water and air. Since 2008 it has also had responsibility for all flood and coastal erosion risk management. Under the current planning system local authorities must now consult the Environment Agency and also a list of other government agencies such as Natural England and non-governmental interest groups when considering planning applications. In this way the disparate initiatives to protect the environment have been integrated with the development control system. In 2012 the government published 'The Natural Choice: securing the value of Nature' the first White Paper on this subject for over 20 years. Its vision is:

'By 2060, our essential natural assets will
be contributing fully to robust and resilient
ecosystems, providing a wide range of goods
and services so that increasing numbers of
people enjoy benefits from a healthier natural
environment.'[90]

This integrated approach to environmental protection has matured into a concern
for sustainable development – the subject of the final section of this chapter.

SUSTAINABLE DEVELOPMENT

The key themes in the growth of 'sustainability' as an international movement
affecting architects are: the protection of the natural environment, scarcity of
finite resources, climate change and global warming. The key driver to current
government policies is the link between our production of carbon dioxide through
all mankind's activities from industry to agriculture – and global warming.

The growth of sustainable development as a concept and policy has taken
place on a world stage and involve global initiatives across nations to address
global problems on a scale never previously envisaged perhaps since the Second
World War. Although, there is only one side in this battle, it would be wrong
to conclude that all nations are working in concert to mitigate the perceived
threat to the global environment. Although consensus is growing, there continue
to be disputes about the science, the evidence, the theory and the future risk.
Many countries maintain that it is against their national interests to invest in
international, cross-border measures to mitigate these risks.

Of course 'sustainable development' has a much wider meaning, including
concepts of social and economic sustainability. The phrase has its origins in 'Our
Common Future', the UN's Brundtland Report of 1987.

The Brundtland Report 1987

In this influential report, the UN tried to reconcile two seemingly incompatible concepts: economic development and environmental conservation.

'...the "environment" is where we live; and "development" is what we all do in attempting to improve our lot within that abode. The two are inseparable.'

They promoted the concept of sustainable development, defined as:

'Sustainable development is development that meets the needs of the present without compromising the ability of future generations to meet their own needs.'

The Brundtland Report established three pillars for sustainable development: economic growth; environmental protection; and lastly social equality across the globe. While the first two have been adopted efforts to achieve the last goal are less obvious. Indeed poorer countries see restrictions on industrial activity and hence economic growth as perpetuating global inequality.

The Rio Conference in 1992 was the first major platform for the environment. Word leaders came together to tackle climate change and falling levels of biodiversity. In June 2012 the 'Rio+20 Earth Summit' (the third UN conference on Sustainable Development) took place in Brazil. At the conference the main subject was Brundtland's third pillar, social equality across the globe. The key themes were food, water and energy that were meant to translate into actions to establish a clear path to a sustainable future. Many world leaders stayed away – attending the G10 economic conference instead. The WWF described it as a 'chance squandered'.[91]

This discussion does not intend to stray into the complex science of global ecology that underpins concern for environmental protection and sustainable development. However two powerful themes have driven the initiative. First, the idea of a vulnerable natural world in all its precious biodiversity under threat – exemplified by the destruction of the rainforests and wildlife. Second, the link between carbon dioxide levels in the atmosphere produced by mankind's activities – the so-called anthropogenic causes – on the planet and global warming. The most obvious signs of which are melting ice-caps and changes in the world's weather systems. Both strands have been publicised and popularised. For example, the WWF, with its giant panda logo has transformed itself from a British charity concerned with protecting endangered species – to the world's leading independent conservation organisation: working with governments, businesses and communities.[92]

Global warming has been taken up by pressure groups such as Greenpeace and Friends of the Earth. The Green Party now has an effective, if small, political presence in some western parliaments. The idea of a carbon footprint, whilst ignoring some of the other causes of global warming such as methane gas, is a simple and effective measure of pollution – and social inequality.

Environmental Protection: the World Wildlife Fund (WWF)

The World Wildlife Fund (WWF) – an independent charity set up to conserve the natural world – employs over 300 people and operates in over 100 countries. It has become a major pressure group working under three key principles: conservation, climate change and sustainability.[93] It started in 1961 with the aim of protecting rhino and elephant herds in Africa but over the last fifty years it has moved from a body set up with the aim of protecting species to conservation of both animals and habitat – protecting the interconnected eco-systems. This has developed into a wider interest into the effects of climate change on the environment and the sustainability through changing the way we live and reducing the disproportionate impact developed nations have on the rest of the world. Its press centre in mid-2012 was lobbying on a range of topics from wild tigers to the decline in biodiversity and the green economy.

It has a truly global vision, a strong image and a knack for publicising causes. Its transformation from a single issue wildlife protection group to a global environmental pressure group mirrors the change in attitudes towards the environment over the last fifty years.

In its 2012 Living Planet Report it found that:

- The global Living Planet Index has declined by up to 30 per cent since 1970.

- It is currently taking 1.5 years for the Earth to absorb the CO2 produced and regenerate the renewable resources that people use within one year.

- 2.7 billion people live in areas that experience severe water shortages for at least one month of the year.

- The per capita Ecological Footprint of a high income country such as the USA is currently six times greater than that of a low income country such as Indonesia.

- The UK has risen four places from 31st to 27th place in the report's global consumption ranking, which compares the Ecological Footprint per person, per country.

- The top 10 countries with the biggest Ecological Footprint per person are: Qatar, Kuwait, United Arab Emirates, Denmark, United States of America, Belgium, Australia, Canada, Netherlands and Ireland.

According to the global Living Planet Index, 'declines in biodiversity are highest in low income countries, demonstrating how the poorest and most vulnerable nations are suffering the impacts of the lifestyles of wealthier countries.'

We also continue to consume fifty percent more natural resources than the planet can sustainably produce.[94]

THE UK'S RESPONSE

The UK government shares the belief that commonly agreed goals on the
environment can be achieved through the current economic system:

'Sustainable development does not mean having
less economic development: on the contrary,
a healthy economy is better able to generate
the resources to meet people's needs, and new
investment and environmental improvement often
go hand in hand.'

This sums up the belief shared by the main political parties that the market
economy will meet the objectives of sustainable development. The mechanism by
which this works is not clear and the research for the WWF's Living Planet Report
for 2012 suggests that more radical measures are required.

The responsibility for sustainable development is currently shared by two
government departments: the Department of Energy and Climate Change,
which controls the supply of energy and the Department for Communities and
Local Government which controls sustainable development through the planning
system. It sets performance targets for energy consumption – and therefore
the production of carbon dioxide from buildings in use – through the system of
building regulations (see Chapter 9).

The National Planning Policy Framework introduced in 2012 has tried to simplify
the planning system in order to support and stimulate economic growth while
making sustainable development a cornerstone of government policy. The system
still relies upon statutory consultation with bodies such as the Environment
Agency to integrate national policy objectives with local needs. It has been left to
the local authorities and their planners to control development and at the same
time protect the environment for future generations.

CONCLUSIONS

The UK has developed a series of distinct systems to respond to the separate
concerns for conservation and national heritage, environmental protection and
sustainable development. Some of these are a gradual response to changing
attitudes in society – such as the protection of ancient monuments, the
conservation of individual buildings and the historic urban grain. Others, such as
the control of waste and pollution have been accelerated by European legislation
requiring Environmental Impact Assessments for certain types of development
which have been grafted on to the plan-based system of development control.
The greatest current and future challenges concern sustainable development:
a response, which some say is too little too late, to a global problem requiring
immediate remedial and preventative action as well as changes to economic
and behaviour to protect the environment in the future as well as respond to

current needs. The current system of environmental control, whilst not appearing proactive, does have the potential to accommodate radical change at a local level if the government has the political will to try and implement policies that will both reduce our carbon footprint and protect the environment for future generations.

It has been argued that the political conditions for achieving sustainable development are unachievable.

'It seems far more likely that the relentless process of market liberalisation, dominance of multi-nationals and the pursuit of national interest will accelerate the depletion of resources and the degeneration of the environment.'[95]

As architects of future built environments, you will need to make strategic design decisions about location, materials and building performance which will have long-term effects on our carbon footprint, the use of scarce resources and future well-being. This is a challenge that previous generations of designers – brought up in a world of cheap energy and unquestioned economic growth – did not have to face and will in turn produce new forms of architecture.

FURTHER READING

Town Planning

The standard text for urban designers on this subject is 'Town and Country Planning in the UK' (14th ed.) by Cullingworth B and Nadin V Routledge 2006

Heritage and Conservation

From a planning perspective 'British Planning' 50 years of Urban and Regional Policy' is a collection of essays on a number of subjects including a chapter on Heritage by Peter Larkham which is often quoted elsewhere. (Cullingworth B (ed.)) Athlone Press 1999

Sustainability

www.wwf.org.uk

9
PLAYING SAFE
– BUILDING RESPONSIBLY

INTRODUCTION

This chapter discusses how safety has become an important factor in building design, in construction and in use.

Construction is a relatively dangerous activity compared with other industries such as manufacturing. Construction sites, in particular, are dangerous places with a disproportionate number of accidents – some of which are fatal. Although the construction industry accounts for only about 5% of the employees in Britain it still accounts for 27% of fatal injuries and 9% of reported major industries. In 2010 there were 50 fatal injuries and 2.3 million work days were lost due to work-related illness or injury. Although these figures are bad they show an improvement over previous years and are part of a long-term trend of safety improvements in the industry. For example, in 1990 there were 120 fatalities.[96]

Completed buildings ought to be less dangerous but the effects of poor design, specification, and construction, or events such as fire, can be catastrophic. Although uncommon, events like these do happen. This is unacceptable: users should be able to expect that every building has been professionally designed to be safe and constructed competently using appropriate materials and equipment.

This chapter looks first at how the rules governing the design of buildings have developed and been codified – principally through the creation of the Building Regulations. It then looks at how, more recently, as a result of wider European legislation, the inherent dangers of construction processes have been regulated by the Construction Design and Management Regulations – not through rules but by setting out responsibilities and management processes with the aim of improving safety on construction sites and buildings in use. Both involve different attitudes to risk. The first sets out rules which if kept to will reduce the risk of something happening, such as structural failure. The second sets out a process for managing and controlling the inherent risks of construction.

REGULATING BUILDING DESIGN AND CONSTRUCTION

Historically the main dangers for domestic buildings were structural collapse due to inadequate design and construction, destruction by fire, and illness – possibly death – caused by pollution and waste. These problems are heightened in the urban setting where the density of building, and hence population, increases the risks. Efforts to control building design and construction as a way of making the urban environment safer pre-date the creation of urban planning and heritage legislation discussed in the earlier chapters. The progress of regulations governing safety and well-being have principally been due to specific responses to particular dangers – usually due to large-scale calamitous events coupled with changes in social attitudes. The level and effectiveness of regulations are also linked to the capacity of government, both central and local, to execute and administer them. Attempts to regulate construction only became effective after the creation of relatively well-organised local governments and resources funded by local taxation in the large metropolises of Victorian Britain.

EARLY ATTEMPTS TO REGULATE DESIGN AND CONSTRUCTION

The early attempts to regulate construction – to apply rules – generally failed because the structure to enforce the rules was either absent or ineffective. Without an effective central or local government system to administer and police rules they generally failed to be applied and without effective sanctions they were often ignored. Where rudimentary administrative structures did exist officials in power, without sufficient accountability, could often be persuaded to 'turn a blind eye'.

One of the earliest examples of attempted control was in seventeenth century London. At a time when drainage was a novel idea all waste was simply dumped in the street. Although the science linking environmental factors with health was not developed, the link between appalling living conditions and bubonic plague was understood – if only by association. In 1603, a serious outbreak killed 30,000 people in London alone. A further outbreak in 1665 killed 80,000 – one in six of London's inhabitants. The simplest way of preventing the build-up of waste – from humans, animals, food and processes such as tanning was to restrict or stop new development.

By the end of Queen Elizabeth I's reign in 1603 the enforcement of existing rules barring any building on new foundations were tightened up.

In March 1605 the new king, King James I, issued a ban on all new building for six months. New standards of construction such as a minimum building width of at least twenty feet (six metres) and ceiling heights of ten feet (three metres) for new construction were specified.

In 1607 a further royal proclamation controlled new building around London by requiring a special licence to be obtained.

Although there were several prosecutions with penalties including the demolition of properties, these were few and far between. The wealthy could simply buy an exemption from the Exchequer. The revenue from these fees became a useful source of funds and so the proclamation became less effective as a way of restricting urban growth.

In 1615 a Commission for Building was established to regulate building in London. This included a bar on overhangs, bay windows and thatched roofs, and a stipulation for uniform building widths with minimum wall thicknesses.[97]

The Great Fire of London in 1666 destroyed a major part of the historic city and led to major changes in the way buildings were designed and constructed. There had been previous major fires – the last in 1635 – and the fire risks caused by poor construction, 'jettied' floors, timber construction and thatched roofs were well known. One of the major characteristics of the Great Fire was the speed at which it spread from building to building in the densely built-up and populated city centre. The existing rules had not been applied effectively. Proclamations from Charles II in 1661 and 1665 had little effect. The Fire devastated a large area of the financial heart of the capital. National pride was also at stake because the Great Fire became international news. As a result attitudes to enforcement changed. The main driver for change was not personal safety but the protection of property. The existing regulations designed to prevent the spread of fire from one property to another were tightened up and fully enforced. This involved further restrictions on both design and materials. These changes transformed the city and led to the uniformity and integrity that we recognise in Georgian architecture. The use of inflammable materials such as timber and thatch were prohibited – and the rules were enforced. Projections in front of the building line such as the large overhanging timber eaves and the 'jettied' overhangs that characterised half-timbered construction were also prohibited.[98] Windows had to be set behind the building line rather than flush or projecting. Brick was specified as the main material for façades, with the roof set behind a parapet. One regulation that has remained and even survived metrication in 1971 relates to the position of windows. Under the London Building Acts all windows above ground level had to be a minimum of three feet (now one metre) from the boundary line.

PUBLIC HEALTH AND SAFETY: THE GROWTH OF LOCAL GOVERNMENT

The rapid industrial growth of cities and towns in the nineteenth century required the equally rapid provision of new, larger industrial buildings and densely-packed housing to accommodate the growing working class. Unrestricted urban growth and a lack of building standards led to poorly constructed housing and sudden structural collapse.

The growth of industry led to toxic air and water pollution. Densely-packed housing close to industrial sites and the lack of effective sanitation led to public health problems and disease. It took some time to establish the relationship between poor housing, inadequate sanitation, pollution and public health. The 'Great Stink' in London and further major epidemics led to changes in society's attitudes. It is interesting that the need to enforce rules to deal with the adverse effects of the growth of towns and cities also led to the creation of effective local government. As well as funding or facilitating infrastructure projects local authorities began to impose basic standards of construction through local laws or 'bye-laws' as they were called. The legacy of these basic rules for construction standards can be seen in the uniform rows of Victorian 'bye-law' terraced housing built with effective sanitation and to a safe standard of construction.

Local bye-laws made and administered by individual local authorities under successive Public Health Acts regulated the design and construction of buildings until 1965. There were local variations but increasingly the regulations were based on model bye-laws issued by central government. In 1965 the first set of Building Regulations was issued and the Regulations were amended regularly until the current Building Act of 1984. The current Building Regulations are shorter and less prescriptive. The intention is that regulations can be interpreted to achieve reasonable standards of health and safety in design and construction.

CURRENT LEGISLATION

The English (and Scottish) systems have evolved into a comprehensive regulatory regime. In England the underlying legislation is the Building Act 1984. This consolidated earlier legislation and also brought London into the national regulatory regime (although certain parts of the London Building Acts, principally concerned with fire safety in tall buildings, remain as additional requirements).

The Building Regulations sit below the underlying Building Act 1984 allowing for technical changes to be made without going back to Parliament. For example the Building Regulations 2000 were revoked in 2010 and have since been amended by the Building (Amendment) Regulations 2011.

The Building Regulations are in two parts:

1.   Procedural regulations that set out what kind of work needs approval,

and

2.   Technical requirements that set the acceptable standards for building work.[99]

The procedural regulations explain how to obtain approval. Although the Regulations are published by the Department for Communities and Local

Government the local authorities are generally responsible for administering the application and technical approval system. For more information on the detail of the approval system you can look at any local authority website. (In Scotland the approval is through the 'Building Warrant' and slightly different procedures apply.)

The approval system was originally administered only by local authority Building Control Officers. (In London the old title District Surveyor still remains in some boroughs.) However the Building Act 1984 introduced 'Approved Inspectors' and led the way for the private sector to also administer the system. This approval route was slow to take off, principally due to concerns about liability following some high-profile court cases involving local authority building control officers.[100] The responsibility for checking compliance now rests with BCBs (Building Control Bodies). These include local authorities and also private sector consultants who specialise in Building Regulations approval. The advantages of private sector consultants is that they can be appointed early in the design process and contribute to the detailed design and enter into a dialogue with the design team evaluating and commenting on alternative technical design solutions rather than acting purely as an approval body.

The technical requirements that set down acceptable standards for design and construction are to be found in the current set of Approved Documents. The advantage of this system is that detailed technical standards can be updated quickly.

The Approved Documents in 2012 were:

A   (Structural Safety)

B   (Fire Safety)

C   (Resistance to Contamination and Moisture)

D   (Toxic Substances)

E   (Resistance to Sound)

F   (Ventilation)

G   (Sanitation, Hot Water Safety and Water Efficiency)

H   (Drainage and Waste Disposal)

J   (Heat-producing Appliances)

K   (Protection from Falling)

L   (Conservation of Fuel and Power)

M   (Access to and Use of Buildings)

N   (Glazing Safety)

P   (Electrical Safety)

Workmanship and Materials[101]

(note: the detail of the ADs continue to evolve)

COMMENTARY

With the exception of Approved Documents E and L, all the rest are concerned with safety. It is also interesting to see that the list generally follows the chronology of the evolution of the regulations starting with A (Structural Safety) and B (Fire Safety). Approved Document P (Electrical Safety) was added late because the government and local authorities, although realising that electrical safety was paramount, were reluctant to get involved with the approval of electrical installations, primarily because they did not possess the expertise. Building owners relied instead on a separate set of industry-approved regulations (the IEE (Institute of Electrical Engineers) Regulations). Compliance with AD P is now demonstrated by a self-certification process – in effect bringing the industry approval system into the statutory approval system. This is an example of a more pragmatic approach to safety, regulation and approval.

If you look at the Approved Documents you will see that the relevant part of the Building Act is quoted and the Approved Document in effect shows how to comply with the Building Regulations. However, the guidance specifically does not constrain designers. They have the flexibility to develop alternative design solutions. These will be acceptable if it can then be shown that they comply with the Building Regulations.

A CASCADE OF INFORMATION

The cascade of information starting with the Building Act 1984, the Building Regulations (routinely updated) and the Approved Documents sit below these. Sitting below the Approved Documents are Approved Codes of Practice, British Standards and European Technical Standards. (See figure 9.1)

To see how this cascade works you can look at one example: Approved Document B. This is split into Volumes 1 and 2. Volume 1 applies to dwellings and Volume 2 applies to any other building such as offices or shops.

Approved Document B sets down factors that affect fire strategy such as the maximum distance you can reasonably be expected to travel in an emergency to reach a safe place. You can see that this will affect the size and shape of the floorplate of buildings such as offices as well as staircases and fire exits.

Maintaining standards through legislation immediately requires that standards are defined, which quickly makes the overall picture much more complex. A key part of safety, as you have seen in response to the Great Fire of London, is the properties of materials when subjected to heat. It would be pointless to have a fire exit that is lined with flammable materials or is constructed of materials that do not resist fire well. It is important therefore, that these materials meet certain standards – and that involves testing in a standardised way. In the case of weak points on fire exit routes – fire doors for instance – these need to be tested as well to see if the *combination* of materials: glass, wood, even the timber beads that hold the glass in place and the hinges that hold the doors will resist fire as well.

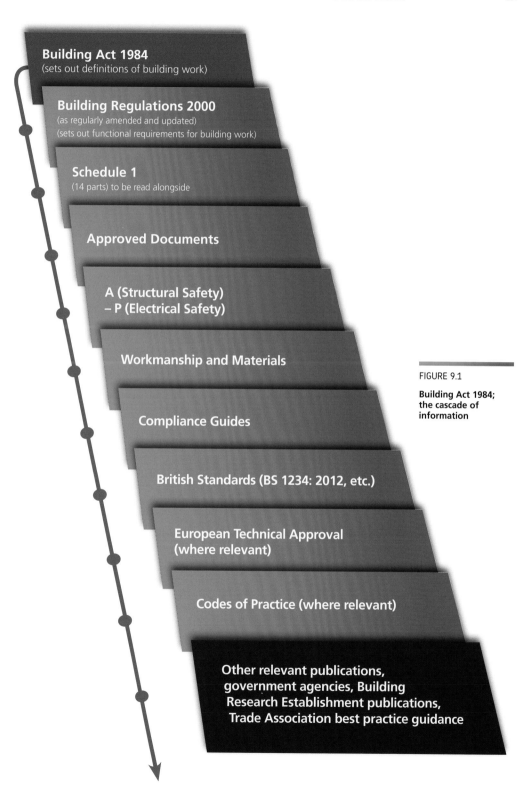

Building Act 1984
(sets out definitions of building work)

Building Regulations 2000
(as regularly amended and updated)
(sets out functional requirements for building work)

Schedule 1
(14 parts) to be read alongside

Approved Documents

**A (Structural Safety)
– P (Electrical Safety)**

Workmanship and Materials

Compliance Guides

British Standards (BS 1234: 2012, etc.)

**European Technical Approval
(where relevant)**

Codes of Practice (where relevant)

**Other relevant publications,
government agencies, Building
Research Establishment publications,
Trade Association best practice guidance**

FIGURE 9.1

**Building Act 1984;
the cascade of
information**

A key part of Approved Document B is the tests that establish:

a)   the resistance of materials to fire; and

b)   the flammability of the surface (surface spread of flame).

The set of tests for building materials is not found in the Approved Document but in a British Standard: BS 476 – the next step in the cascade. (The BS itself has its own cascade: it is split into a number of parts each detailing different testing procedures). Therefore if a designer specifies a material or a door design that passes the test then it will be satisfactory for the purposes of Approved Document B.

One of the effects of the testing procedure is that if you design something new or use materials in a different way they will have to be tested. This also applies to goods and materials sourced from outside the UK. Unfortunately this can be seen as a barrier to trade – especially within the European Union where there are not meant to be any trading restrictions.

THE IMPACT OF EUROPEAN LEGISLATION: AN EXAMPLE

The Single Market requires that barriers to trade are removed, a prerequisite that gave rise to the European-wide Construction Products Directive (CPD). In the construction industry, this led to the UK Construction Products Regulations (CPR). The testing regime set down in BS476 presented a problem. As well as meeting UK tests it could lead to another tier of tests to show that products met European regulations and could be traded freely in other EU states. The CE marking, showing compliance with EU technical standards is, in effect, a 'passport' for goods within the EU. The problem for the UK was materials tested in accordance with BS 476 did not comply with the new European tests. Worse still, the European tests used a different methodology and were tougher than those set down in BS 476. Materials that gave a 60 minute fire resistance under BS476 were now found to have a resistance of approximately 50 minutes using the European tests, meaning that materials that used to meet the CPR no longer did.

This had the potential to place an unreasonable burden on manufacturers who had relied on BS476, and on the professionals who had got used to specifying to that standard. To assess the extent of the burden, the government used a procedure called a Regulatory Impact Assessment (RIA). RIAs were introduced in 1998 with the objective of cutting the amount of 'red tape' in government. The aim was to strike a balance between under-regulating – which might affect public safety – and over-regulation which might produce too great a burden of bureaucracy on the industry. There is an example of this policy in action in Approved Document 'B' – which also shows how the government responds to European legislation at a technical level. The RIA looked at different options and the costs of implementation. The result was that as the UK had implemented the CPR we had to comply with the new testing regime anyway and a new European supplement to Approved Document B had to be written. It was estimated that the cost to the construction industry to familiarise itself with the new regulations alone would be £3.5m.[102] This excluded the cost of changes to the tests themselves and re-engineering materials and designs to meet the new regulations. The government accepted that without these changes, and the CE label showing compliance, the UK would not be able to sell goods on to other EU states.

As a result, a new supplement to Approved Document B was produced. One of the important conclusions from this exercise is that architects must first be aware of the cascade of information, second the content and third to be alert to changes in the system of regulation.

REGULATING FUTURE DESIGN AND CONSTRUCTION

The Building Act 1984 created two key changes to the regulatory system. First, it introduced the idea of 'Approved Inspectors', independent of the public sector and responsible for approving design and construction. As local authorities have become more enterprising and seek other sources of revenue they too are offering their services outside their geographical areas. The local building control department is now one of a number of BCBs (Building Control Bodies). Second, it created the Approved Documents which could be updated and expanded to suit changes in current practice, government legislation and European legislation. They were also meant to encourage innovation rather than slavish adherence to rigid rules. Although there have been some successes, including new approaches to fire safety in relation to large buildings and atrium designs that objective has not really worked.

The next section shows how the regulatory framework has changed still further.

HEALTH & SAFETY LEGISLATION: THE HEALTH AND SAFETY AT WORK ACT 1974

The Health and Safety at Work Act 1974 (HSAW Act) is the primary piece of legislation governing safety in the workplace. Construction sites therefore are subject to the requirements of the HSAW Act. The HSAW Act itself is interesting because it grew out of a report commissioned by the government of the day into factory accidents. It was led by a former chairman of the then nationalised coal industry, the National Coal Board – appropriate given the poor safety record of coal mining. The purpose of the legislation is to reduce physical injury and harm. This may seem obvious but you will also see that there is an overlap with the common law area of tort and negligence in particular (discussed in Chapter 6). The important point about negligence is that it sits outside contract law and, in relation to this legislation, follows two key principles: the duty of care we owe to others (the 'Good neighbour' rule) and the link between a failure to exercise that duty of care and the physical injury or harm that is caused by that failure.[103] It also sets out principles for compensation (or damages). The HSAW Act takes these common law principles and applies them to the workplace. The following gives a good example of the H&SAW Act in action.

Ikea fined £75,000 after boy's fingers are sliced off in escalator

'The two-year old boy was holding his father's hand when he lost his balance on the downward slope of the travelator. As he fell his hand became trapped between the skirting and the pallet treads. The gap was bigger than the legal limit of four millimetres.

Ikea initially denied breaching the HSAW Act at the store in Milton Keynes on the basis that Lift Serve, a maintenance company, conducted tests two days before the accident in 2007 stating that the skirting and tread were safe. However, it has now accepted both criminal and civil liability and is paying compensation to the boy, who is now seven.

At the sentencing hearing at Aylesbury Crown Court the judge fined the company £75,000 and ordered it to pay £90,000 in costs.

Barry Berlin, prosecuting, said Ikea had failed to compile a risk assessment of the travelator and check that maintenance had been carried out.

Simon Antrobus, defending, said Ikea regretted the incident but stressed that civil proceedings had found Lift Serve 75% liable and Ikea 25%.

Judge Cutts said: 'There is no doubt in my mind that Ikea, in this store and on this particular travelator exposed children to severe and permanent disability. Ikea had an indelible duty to ensure the safety of all customers in the store and failed in its duty here.'[104]

COMMENTARY

The report of this tragic accident raises some interesting points. First, you will see that it took five years for the case to reach the court – a very long time. This may partly be due to the second point of interest – a common law civil action for negligence where the liability was shared 75:25 between Lift Serve and Ikea. Ikea may have asked for the criminal case to be delayed until the common law civil case had been settled. As it originally denied liability it may have been hoping that Lift Serve would have been found 100% liable. Damages in tort, although not referred to, would have been paid in this proportion (75:25) by the two companies to the seven-year old. Third, the case shows that the common law duty of care becomes a statutory duty in the workplace: Ikea could not contract out of its statutory duty even though it had employed Lift Serve to maintain and test the travelator. It also owed a duty of care to anyone using it in the store – not just its customers. Fourth, as the judge said, Ikea had 'an indelible duty'. Fifth, Mr Berlin refers to the failure to carry out a risk assessment – a risk management process embedded in the health and safety regulations. Lastly, it is worth emphasising that Ikea was *prosecuted* because a breach of the statutory duty is a *crime* and punishable by a fine or imprisonment.

SAFER BUILDING: THE CONSTRUCTION (DESIGN AND MANAGEMENT) REGULATIONS 2007

The example above shows that the application of HSAW Act is far-reaching. The Act has therefore always applied to construction sites as well as the workplaces and buildings that architects design and contractors build. However a separate set of regulations now apply to the construction industry: the Construction (Design and Management) Regulations 2007 (CDM Regulations). The CDM Regulations sit within the HSAW Act and owe their origin to a 1992 European Directive.[105] (The HSAW Act was a convenient and appropriate place to locate them without introducing new primary legislation.) The key point is that the CDM Regulations place legal duties on virtually *everyone* involved in construction work.

The CDM Regulations are also interesting in that they represent a shift in government and industry thinking away from a rule-based system to a procedural management system. The regulations are overseen by the Health and Safety Executive (HSE), the government body responsible for all health and safety legislation in England and Wales. Rules have been replaced by responsibilities and procedures. Unlike the Building Regulations where you submit drawings and calculations for approval by a BCB, the system is intended to be largely self-regulating. Therefore you can be prosecuted for failing to follow the procedures or the duties required by the CDM Regulations even if an accident has not taken place. Although it is more common for contractors to be caught out, architects have been prosecuted for not following the procedures so they need to be aware of their duties.

Following the principles of negligence law, the CDM Regulations stipulate the duties set out in the HSAW Act and place them in the more fluid context of construction work. The Regulations set out responsibilities for all members of the project team: client, designers and contractors. As the name suggests it requires the systematic management of health and safety procedures in construction projects but, following the principles of negligence law, you cannot contract out of the duties set down in the CDM Regulations. This is important because the tone of the CDM Regulations 'feels' like a management process where different members of the team are contracted to carry out specific work: design, management and construction. Even if the function is contracted out – and on most projects the CDM Regulations require the appointment of a specialist CDM Co-ordinator trained for the job – the fundamental statutory duty of care remains with the key stakeholders: the client, the design team and all contractors involved with the project. Each member of the team has a stake in the health and safety of the project – regardless of their contractual position. Because health and safety is so important it is a mandatory element of annual CPD (Continuous Professional Development) for Chartered architects required by the RIBA.

The next two sections explain first the duties of key stakeholders or 'dutyholders' and the key management process, risk assessment – used to discharge those duties.

KEY HEALTH AND SAFETY DUTYHOLDERS IN CONSTRUCTION PROJECTS

The CDM Regulations specifically refers to six groups of dutyholders: clients, CDM co-ordinators, designers, principal contractors, other contractors and construction workers.[106] Each dutyholder makes a contribution to the safety of construction sites and construction work. They also contribute to the safety of buildings once they have been completed – from maintenance to demolition. Architects and the rest of the design team therefore have to consider how buildings are built and then maintained: a 'cradle-to-grave' approach.

The HSE acts as both an advisor and a monitor for Health and Safety across all sectors, not only construction. It disseminates 'best practice' guidance and acts as a positive force for health and safety practice. It also acts as enforcer and prosecutor – when serious accidents occur.

As a government department that is the independent 'watchdog' for all work-related health and safety, the HSE regulates construction health and safety by monitoring all construction sites. Therefore there is a duty to notify the HSE of construction work taking place. However because construction activity is so diverse and can mean anything from a small house extension to a sports stadium the HSE has to balance its monitoring role with available resources. Currently certain small works need not be notified to the HSE and domestic clients are exempt from all duties under the Regulations. (The common law duties remain, though.) These relaxations of the CDM Regulations are being challenged by the EC as an inadequate implementation of the EC Directive 92/57/EEC and may be removed in due course.[107]

MANAGING RISKS

As architects, it is important to understand and bear in mind the risks that might arise out of your designs. Construction sites are dangerous places. Dangers include working at height, working below ground level, working with heavy and/ or bulky materials and working with portable and fixed power tools. Temporary works such as scaffolding or shoring in a deep trench also present particular hazards.

A vast amount of construction activity concerns buildings that already exist – we only add approximately two per cent to the built environment per year and existing buildings have their own hazards. As well as potential structural problems and the failure of materials such as roof coverings with age it is very likely that any building built before 1990 will have asbestos in it somewhere – either as insulation, as a building material in rigid panels or even in textured paint. Asbestos is particularly dangerous as there is no recognised 'safe' limit to exposure and the effects may take many years to appear. Much old paint also contains potentially poisonous lead. These materials remain safe until you start to alter the building.

When the building is complete a new set of stakeholders will take over. During the life of a building it will have to be maintained. Typically, mechanical plant, such as air-conditioning equipment has a much shorter life than the building itself and will need to be replaced. Windows will need to be cleaned and maintained and therefore need to be accessible. All these aspects need to be considered at the design stage.

HOW DO YOU PLAN FOR THESE UNKNOWNS? HOW DO YOU DESIGN WITH SAFETY IN MIND?

The recognised method for evaluating hazards in construction is called Risk Assessment. This is a general business management tool designed to identify and evaluate possible risks in an uncertain business environment. It is a step-by-step process that identifies events and then evaluates them in terms of likelihood (probability of something occurring) and impact (the likely effect of the occurrence). Having identified and evaluated risks you can then apply appropriate risk strategies. The method is particularly applicable to hazards on construction sites and the HSE has published a standard methodology which is set out below.

Risk Assessment methodology:

1. Identify the hazards.

2. Decide who might be harmed and how.

3. Evaluate the risks and decide on precaution.

4. Record your findings and implement them.

5. Review your assessment and update if necessary.

To help risk assessment by designers the HSE publishes a useful set of lists of what to avoid, eliminate or encourage. These lists are colour-coded red, amber and green.

■ Red List items are to be eliminated where possible; for example: the specification of fragile rooflights and roof assemblies and 'glazing that cannot be accessed safely'.

■ Amber List items are to be eliminated or reduced as far as possible and only specified where unavoidable; for example, 'Specification of large and heavy glass panels'; and 'Specification of curtain walling systems without the provision for tying-in of temporary supports such as scaffolding'.

■ Green list items are that should be encouraged; for example, 'practical and safe methods of window cleaning' and 'off-site fabrication and prefabricated elements to minimise on site hazards.'

There are hazards in maintenance too. Some risk is acceptable, provided it can be carefully managed. Risk assessment can be a design opportunity if integrated into the design early enough. For example, 1 St Marys Axe ('The Gherkin') designed by Foster & Partners in the City of London, is very tall and constructed using a skin of curved glass elements. While the shape and glazing are fundamental to the success of the design concept, working out how to clean the windows presented significant health and safety risks. Luckily the risks were identified and an elegant, integrated solution, a stainless steel ring at high level, was found. It is a slight compromise but barely visible and certainly it has not affected the basic concept.

The failure to consider risks during design stages can lead to bulky, 'bolted on', 'engineered' solutions that destroy the architect's design. Paying heed to the hazards and risks in construction should feature as highly as other physical environmental factors such as the limits of the site. Your building design will be safe to build and use, and you will be complying with the law.

CONCLUSIONS

In this chapter we saw the shift in the way we regulate the design and construction of buildings with public health and safety in mind. What started as a rule-based, prescriptive system where following the rules showed compliance with legal standards has been transformed into a management approach to meet specific statutory objectives. This is part of a deliberate regulatory shift by government: moving the responsibility for compliance from the state to the individual or organisation. The Building Act 1984, while showing ways to comply with the legislation through the use of the Approved Documents, also provides the opportunity for designers and product manufacturers to innovate and provide alternative solutions. The new system of BCBs has encouraged the private sector to provide compliance services – shifting the burden away from local government and providing the resources to manage much larger projects more effectively.

The CDM Regulations takes this process a step further.

Mirroring our common law duty of care in the tort of negligence, health and safety legislation makes this duty of care statutory through the creation of 'dutyholders' – which includes almost everyone involved in the construction process. Risk assessment is used to manage construction risk and improve the health and safety not only on construction sites but also of buildings in use. This is far more effective than trying to write a comprehensive set of preventative rules for every eventuality, even if it was possible. Compliance with the law can be monitored and where failures occur accidents can be investigated, dutyholders prosecuted and lessons learnt.

This chapter has not examined every piece of health and safety legislation. For example, fire regulations which apply to buildings in use have not been discussed. However it is worth noting that this area of health and safety legislation has been updated, moving away from set rules and 'Fire Certificates' issued by the local fire authority to a 'risk-based' approach where buildings are assessed for fire risk. The Regulatory Reform (Fire Safety) Order 2005 sets out responsibilities for fire safety and places the responsibility for the safety of employees and building users with business owners. Finally, set rules will remain where they are helpful and informative. It is likely that the approach to health and safety and risk where the responsibilities to society are based on common law principles of our duty of care to others will be the template for the future, rather than relying on the state to set down rules and then police them.

FURTHER READING

In researching this subject it is better to go straight to the original sources available on the government websites:

Building Regulations

www.planningportal.gov.uk

www.communities.gov.uk

CDM Regulations

The HSE website is the best source of information for all things to do with CDM. In particular the Approved Code of Practice (ACOP) is required reading for architects and designers in practice

www.hse.gov.uk

Other helpful guides include:

Managing health and safety in construction

The Absolutely Essential Health and Safety Toolkit

(Available on the HSE website)

These sources also include the procedural aspects of statutory compliance which are not covered in this book but are essential for the architect in practice.

10
BEYOND THE STUDIO
– MAKING IT HAPPEN

INTRODUCTION

Procurement is the process by which your design is transformed into a built form. This transformation requires the involvement of many different stakeholders – some active and some passive, some directly and others indirectly involved.

The **construction contract** is the legal agreement – the physical document (if there is one) – that flows from the procurement decisions made by stakeholders. It is an agreement made between the key stakeholders that allocates risks as well as benefits and responsibilities – and is legally enforceable.

Your design work is integral, but may not necessarily be the starting point, to the procurement process and the design will change and adapt to accommodate the requirements and decisions of stakeholders – from outline design to construction.

Construction is always a complex process – even with small-scale projects. This complexity and the 'fluidity' of the outcome – you do not really know what will be built until it is complete – leads to uncertainty and risks. These can be design risks that affect the technical performance of building elements or the entire building that flow from a new design concept or experimental use materials. There may be technical risks that affect the construction process and possibly time and/or construction cost. Location and even access will all pose logistical problems that in turn may affect time and cost.

The complexity, scale and cost of a design project together with the attitudes of major stakeholders to design and construction risks all affect the way it is procured and built.

This chapter starts by considering stakeholders and their attitudes to risk and complexity. It then gives an overview of the construction industry and how procurement has changed in recent times. This is followed by an explanation of the key methods of procurement and a brief description of the different standard forms of contract.

STAKEHOLDERS: PEOPLE AND PROJECTS

The idea of 'stakeholders' is a good starting point for analysing who is involved in your project. Stakeholders can be defined as people who have an interest in your project or are affected by the project. They can be divided into direct stakeholders and indirect stakeholders. Some are passive, but influential, and some are active. Some may initiate projects, be proactive and a positive force whereas some may obstruct or block the project. Your design projects will typically be ambitious in both scale and complexity with a brief that may challenge accepted values. You are likely to have multiple stakeholders and analysing their position and interests will inform both your brief for the project and your design.

For example, your project may involve a re-evaluation of a historic but run-down part of a seaport. The brief could be open-ended: regeneration through new housing or industry or a new cultural centre.

The first step is to map your stakeholders and distinguish between key, direct stakeholders and others.

Some stakeholders may object to the project. For example, public consultation is embedded in the UK town planning system and attempts to block projects are common. Sometimes the process is integral to the briefing process – in projects where public participation is essential.

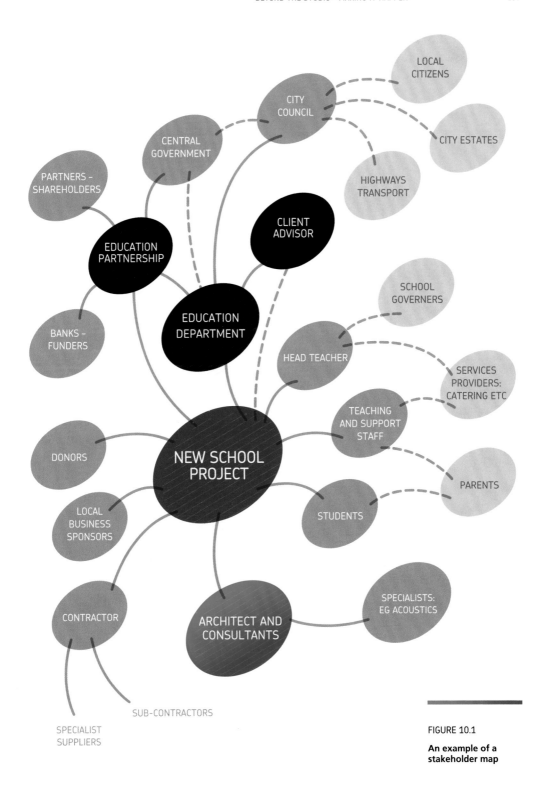

FIGURE 10.1

An example of a stakeholder map

PROCUREMENT OVERVIEW: ATTITUDES TO RISK AND COMPLEXITY

The stakeholders are diverse with many potential interests. Your skill as lead design consultant may be to capture the multiple and conflicting needs of these stakeholders in order to inform and develop the brief. Different procurement methods have been developed to try to manage the different types of project risk. If you can identify key stakeholders' attitudes to particular risks then this will inform how you decide which is the most appropriate procurement route for the project.

STAKEHOLDERS	RISK AVOIDERS: 'RISK AVERSE'	RISK MANAGERS: 'RISK ACCEPTORS'
Financial risk	Funders Banks	Shareholders
	Local government	Commercial developers
	Central government	'Serial' clients: (universities possibly)
	European Union	
	End-user clients (homeowners)	
Design risk	Contractors	Designers
		Engineers
Construction risk		Contractors
(Time, cost and buildability)		Sub-contractors
		Designers, Engineers
(Health and safety)	All stakeholders	All stakeholders
Risks in use	Users	
(Health & Safety)	Maintenance	
Building performance	Designers, engineers	
	Contractors	

FIGURE 10.2

Stakeholders' attitudes to risk

The key stakeholders in deciding the appropriate procurement route are the client and their funders who are concerned with who are with financial risk. From their point of view the main risks are of a project going over budget or being delivered late. However successful the design, exceeding the budget or extending the programme will seriously compromise the success of the project. In the private sector the economic viability will depend primarily on the cost of the project rather than the merits of the design. Funders, including banks and companies as well as individuals, will require a return on their investment and will normally have chosen your project instead of other competing projects on pure financial grounds. If the cost escalates the percentage return on their investment will reduce. Excessive cost over-runs may even jeopardise the completion of a project if funds dry up. In the case of the public sector the project will also have been approved in competition with other projects: the budget will be fixed. Similarly, any delay to the programme in a private sector project will adversely affect the revenue streams – delaying the rental from a tenant or the money from a valuable sale. In the public sector delay may not affect revenue but it will delay the move from other premises and possibly the cost of renting temporary accommodation as well as delaying the delivery of a valuable resource for the community such as a library or leisure centre. A school that does not open at the beginning of the academic year will have huge knock-on effects.

The design team and the construction team may be consulted – but not always. Procurement, therefore, is often a strategic decision taken by key stakeholders before the design team has been appointed. You will also see that attitudes to risk may change according to the phase of the project. For example, every member of the team is involved in managing health and safety risks.

Before considering each procurement route in detail it is worth placing your design work and the decision about procurement in the context of the wider construction industry.

THE CONSTRUCTION INDUSTRY: AN OVERVIEW

In 2012 the UK construction industry represented approximately eight per cent of our national income (Gross Domestic Product) and was a major contributor to economic growth in an otherwise 'flat' year for the British economy.[108] It is also a major employer and exporter of construction expertise – this includes including architectural and structural design consultancy and construction project management. It is important for architects to remember that design is an integral, rather than separate, part of the construction process and therefore the construction industry. If procurement is the process of getting designs built then you should recognise that you are ultimately working as part of a much larger industry and that procurement decisions, how buildings are built, are sometimes taken before the architect is appointed. Because construction is both complex and risky architects may also be appointed to take on different roles within the design and construction team.

Throughout the latter part of the twentieth century the construction industry was under-capitalised and fragmented. Under-capitalisation meant that the industry was unable to invest in the development of skills and expertise or recruit professional managers and as a result had a relatively poorly trained workforce that was badly managed and paid less than their equivalents in other comparable industries, such as engineering. Also, over the same period, the workforce, including construction managers, was employed on a casual basis – moving from site to site as projects start and finish. (This casual workforce even had a name: 'The Lump' and was mostly invisible to the tax authorities.[109]) As a result, skilled staff were always in demand and moved from project to project. The uncertain nature of projects also meant that companies could not offer a secure career path. Expertise was lost when a project reached completion and key staff moved on. Changes in employment law also made it more expensive to recruit and then release staff. Tradesmen were encouraged to become self-employed or set up small companies that could achieve continuity of employment by working for a range of contractors. The end result was a fragmented industry.

To succeed, construction companies have become umbrella organisations that bid for construction projects and then rely upon an army of skilled sub-contractors to carry out the work. At the core of the company is a marketing, cost estimation and project planning organisation supported by skilled construction and design managers who are responsible for project execution. This includes the management of specialist sub-contractors. Even small contractors who traditionally directly employed a number of trades now rely on specialists, for example, excavation, foundations, brickwork, electrical works and decorations. The general contractor has ceased to exist: everyone is a management contractor now.

SUB-CONTRACTORS

This reliance on sub-contractors who bid for separate parts of the project has had a profound effect on the way projects are built and also designed. Work is now organised in 'packages'. Therefore the design and production information for a project has to be organised to match the different sub-contractor packages such as brickwork or windows. Packages may include specialist design input that is beyond the competence of the architect or engineer. For example, most building cladding will require the specialist input of the selected sub-contractor who will carry out the detailed design and construction. Ideally the input of the specialist designer, fabricator and installer should participate in the design development from an early stage. The architect can benefit from the experience of the specialist contractor and the contractor can develop the detailed design in accordance with the design principles and objectives of the architect. The early involvement in the design process also allows design risk to be managed more effectively and ultimately shared or passed to the specialist.

The fragmentation of the construction industry, the lack of continuity where expertise is lost and the reliance on sub-contractors coupled with an obsession with using competitive bidding and bargaining to obtain the lowest cost had three effects. First, every sub-contractor wanted to make a profit – and this led to an increase in bidding costs. Second, contractors used their position as customers to drive down bid costs as a way of increasing profits. This led to an adversarial culture where disputes between contractor and sub-contractor were common. Lastly the fragmentation led to structural inefficiencies – there being little incentive or the capacity to make improvements.

CHANGING THE CONSTRUCTION INDUSTRY

The construction industry still shows many of the characteristics discussed above. However in recent years a number of major changes have taken place. The opening up of the Single European market has led to multi-national ownership of construction companies, bringing much-needed capital to relatively small under-capitalised UK construction companies. The Single Market has also changed the market for specialist design and fabrication companies and widened the market for UK companies in the rest of Europe. This internationalisation of the industry has led to more enlightened management methods and the growth in the importance of building relationships between contractors and sub-contractors. Larger, better-capitalised international contractors are also more able to finance projects and take overall responsibility for project delivery – the design as well as construction.

There has also been a cultural change in the industry brought about, in part, by two influential reviews: first came the 'Latham Report': 'Constructing the Team'[110] published in 1994. Latham, was tasked with reviewing the procurement and contract arrangements in the UK construction industry with the objective of modernising practices and making improvements as well as addressing the adversarial nature of the construction industry. His report, published jointly by the government and the industry made a number of key recommendations – from improving briefing to different methods of procurement and an emphasis on creating long-term partnering arrangements between stakeholders. A second study, the 'Egan Report': 'Constructing Excellence'[111] published in 1998 was the result of the Construction Task force, a group set up by the government under the leadership of Sir John Egan, a former director of Jaguar Cars. It was tasked with looking from a client's perspective at opportunities for improvements in quality and efficiency and to make the industry more responsive to change. Subsequently there have been a number of related studies promoted by the government and supported by the industry.

The Latham Report grew out of the problems faced by the construction industry that were magnified by recession in the early 1990s. He identified and confronted the adversarial nature of the industry and advocated a new approach to procurement which encouraged partnership between stakeholders as well as contractors and sub-contractors. The Egan Report looked at the inefficiencies caused by fragmentation and proposed better management of the sub-contractor supply chain along the lines of the car industry where efficiencies are achieved through the development of long-term relationships between all stakeholders.

The Latham and Egan reports have had lasting effects within the industry through the active participation of industry bodies such as the Construction Industry Board (CIB). Together with a new organisation, the Movement for Innovation (M4I), they published a series of demonstration project studies which showed the application of some of the ideas of 'Rethinking Construction', such as partnering contracts between stakeholders. Since 2004 M4I has operated within 'Constructing Excellence', *a cross sector, cross supply chain, member led organisation operating for the good of the industry and its stakeholders'.[112]* Also, the publication of Key Performance Indicators (KPIs) since 1999 using such critical factors as client satisfaction, project delivery and health and safety has encouraged a fragmented, competitive industry to share 'best practice'.[113]

The legacy of Latham and Egan, coupled with the structural changes brought about by the Single Market and the consolidation and merging of smaller companies has resulted in a better-managed, better-resourced construction industry. There have also been major changes in the way the government and the public sector generally procure projects with a greater emphasis on long-term relationships and partnerships between increasingly sophisticated contractors and increasingly risk-averse public sector clients.

The changes in the construction industry generally have had a number of effects on the sometimes uneasy relationship between architects, the design team and contractors and the way buildings are procured. Architects are used to having an overview of the whole procurement process from inception to completion, leading the design and construction team, advising on contractor selection and administering the construction phase of the project. In the new procurement environment they may take some or none of these roles or be limited to taking a design role.

It is a natural consequence of the increased complexity of construction procurement that roles become more specialised. As a result the traditional 'holistic' role of the architect has, on larger projects especially, become split so that the project team will include other specialists including project managers and contract administrators who are distinct from the design process. Architects may also only be employed for part of the project – either at the beginning for design development or the production information phase whilst the contract administration may be carried out by the contractor. The next section explains the different procurement choices and the effect on the architect's role in more detail.

THE DIFFERENT METHODS OF PROCUREMENT

Procurement currently falls into four main categories:

- General Contracting (or 'Traditional').
- Design and build.
- Management.
- Collaborative.

Each method is now considered in terms of its suitability for particular stakeholders, the risk profile and the different roles of the architect and the design and construction team.

CONSTRUCTION CONTRACTS – SOME DEFINITIONS

'The Employer'

A construction contract is normally between two parties: the 'Employer' and the 'Contractor'. The Employer is normally the architect's *client* who has commissioned the project and for whom the architect and the design team have prepared the design. In most contracts the Employer and the Client is the same person or organisation but this need not be the case. In large organisations the client may be the end-user – a subsidiary company for example, or a school headmaster but the Employer will be the main company or the education authority respectively.

'Design Team'

The design team is limited to the designers and cost consultants who work directly for the client from early on in the project. The team includes the architect, cost consultant, structural engineer, mechanical services engineers, landscape architects, environmental engineers, acoustic consultants and possibly specialist planning and conservation consultants. The team may also include a specialist client adviser and/or project manager. The design team is normally led by the architect who co-ordinates the work of the team. Each specialist professional consultant is normally appointed directly by the client but sometimes the client may require the architect to employ members of the team as sub-consultants.

'Project Team'

The Project Team includes the design team and contractors, either the management contractor who can advise on programme and buildability or specialist contractors who will have a design as well as construction responsibility for an element of the works, such as the cladding.

'Contractor'

This is a general term for anyone who *contracts* to build. The contractor agrees to build the project to the designs prepared by the architect and the design team (Structural engineer, mechanical and electrical services, etc.) There is no single definition: at one extreme the contractor might be a 'one-man band', at the other, a major public company. The contractor manages the equipment, plant, materials and labour on site, pays his suppliers and workmen, and is paid in stages by the client, 'the Employer', for the completed building. He is not responsible for the design – only the end-product.

The term used in most standard forms of contract is 'Contractor'.

'Main Contractor'

This is another term for a contractor but reinforces the point that he is the main point of contact and is responsible for the entire contract between the Employer and the Contractor, or Main Contractor and not just a part of it. It has no special contractual meaning and is sometimes used within the project team to differentiate between the 'Contractor' and sub-contractors.

'Management Contractor'

A management contractor sub-contracts 100% of the works. He is employed purely to manage the process of construction. He is normally an experienced contractor but does not carry out the works. Each item is sub-contracted to Package Contractors.

The management contractor may be seen as a specialist professional consultant and may be appointed to work with the design team.

'Sub-contractor'

The Contractor cannot build the project without a workforce. His skill is to manage the supply of materials and labour. These include sub-contractors with whom he has a separate contract to carry out parts of the project works. Sub-contractors can include specialist groundworkers (responsible for excavation), bricklayers, steelworkers (who construct the steel frame or install beams), electricians, plumbers and decorators. Sub-contractors only work for the Contractor or Main Contractor and do not have any contractual link with the design team or the Employer (the client).

Sub-contractors are sometimes termed 'domestic sub-contractors'. This does not mean that they only work on domestic projects but reinforces the point that they only work for the contractor and not directly for the Employer.

'Package Contractor'

The project is usually broken down into 'packages'. These can be for particular specialist trades such as brickwork which require a 'package' of drawings, schedules and specifications that the specialist contractor tenders on.

Depending on the type of contract the Package Contractor may be employed direct by the Employer or the Contractor.

The package may not be limited to a single trade but for a part of the works: installing partitions, joinery and hardware (locks, handles, etc.). The contractor gives a price for the works and then manages his own team as a separate contract. In this example, he will have to supply and install the partitions, source and supply the doors and frames, and co-ordinate the supply of the hardware. The advantage to the Main or Management Contractor is that a number of separate parts of the project are sub-contracted to one package contractor who is then responsible for co-ordinating this 'contract within a contract'.

Package contractors may also be responsible for the design of specialist items. These include cladding, lifts, escalators and air conditioning – all of which are beyond the design competence of the design team.

The package may extend to complete building systems – such as pre-fabricated modules for whole building systems.

Because the Contractor does not contract to design the work a separate contract is required between the Employer and the Package Contractor for the design element of the works. Otherwise, there would not be any contractual responsibility for this part of the project.

'Works Contractors and Trade Contractors'

In this chapter sub-contractors who carry out specific parts of the project for the main contractor or management contractor are generally referred to as 'Package Contractors'. Package contractors are sometimes referred to as 'Works Contractors' and 'Trade Contractors'. These are contract terms and therefore depend on the contract being used. In principle they function as sub-contractors but the contract relationships may be different. For example, in management contracts package contractors are called 'Works Contractors', whereas in construction management they are called 'Trade Contractors'. In Design and Build contracts and traditional (general contracting) contracts they are simply called 'sub-contractors'.

'Supplier'

A Supplier normally only supplies goods and does not install them. For example, these may be specialist light fittings to be installed by the electrical sub-contractor.

'Purchaser' and 'Supplier'

The JCT Constructing Excellence Contract uses the term 'Purchaser' for the Employer and 'Supplier' to include the contractor. There are contractual reasons for the distinction but the main aim of the contract is to encourage collaboration between stakeholders.

GENERAL CONTRACTING OR 'TRADITIONAL'

General Contracting is the traditional procurement route which has been used successfully for many years. It is the most common method of procurement for projects of all sizes and types. All design work is carried out by the architect and the design team and in turn the design is built by the contractor and his sub-contractors. It sits within a linear model of the procurement process that closely follows the basic RIBA Outline Plan of Work (OPoW). This begins with outline and detailed design, detailed production information, inviting bids for the work based on detailed drawings and specifications, contractor selection and construction.

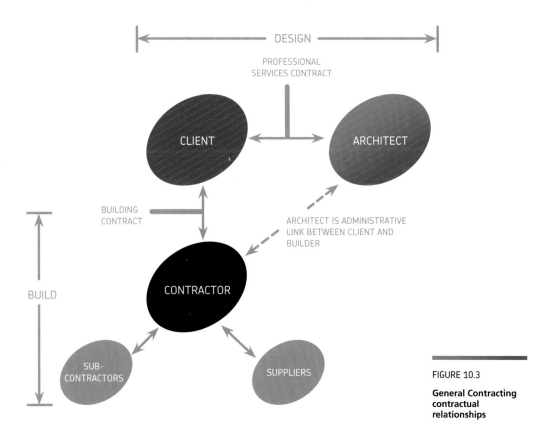

FIGURE 10.3

General Contracting contractual relationships

STAKEHOLDERS

It suits simple stakeholder relationships where there is a single key stakeholder who may often be the key client and end-user: for example, a homeowner or a school that pays for the work and uses the building. These stakeholders are typically risk-averse and conservative in their attitude to both design and construction risks. Traditionally the design risks are managed by relying on the architect to lead both the design and construction processes. The construction risks are managed by the contractor. The architect, as well as leading the design team and the procurement process generally, also acts as the construction contract administrator. Inexperienced clients will rely on the skill and experience of the architect as leader of the design and construction process as they do not usually have the necessary knowledge or resources to effectively manage the project risks.

DISTINCTION BETWEEN DESIGN AND CONSTRUCTION AND THE ROLE OF THE ARCHITECT

In general contracting there is a clear distinction between designing and building. The procurement process assumes that the architect designs and details everything and takes responsibility for the design. Contractors price the work on the basis of the design and other information provided by the client. The design information is passed to the contractor who then builds according to the information provided. The architect is the key point of contact between the project delivery and the client. S/he acts as the link between the demand side (the client) and the supply side (the contractor). The architect will lead the design team co-ordinating the work of other specialists such as the structural engineer, environmental engineer, cost consultant and CDM Co-ordinator.

The architect, in conjunction with the cost consultant, will normally advise the client on the selection of the contractor and then administer the terms of the construction contract. The construction contract is between the 'contractor' and the client ('the employer') but other members of the project team may have roles set out in the contract.

Traditionally the architect acts as the client/employer's agent – issuing instructions, inspecting the works and generally ensuring that the work is carried out in accordance with the design drawings and specifications issued to the contractor. The architect also has an independent role – deciding claims by the contractor for delay or additional costs, authorising interim and final payment to the contractor and stating when the building is complete. In effect the client relies upon the architect to be responsible for the architectural design, manage the design team and procurement process and then the delivery of the entire project.

The main advantage of general contracting is that the architect remains the main point of contact for the client throughout the process. The design risk is managed directly by the architect and the construction risks are the responsibility of the contractor. The distinction between design and construction and the provision of all the design information at the time contractors are bidding for the work, before the contractor is appointed, means that the project should be delivered for the agreed cost in the agreed time. By being in control of the design and administering the construction contract the architect can, in theory, ensure that what was designed is built to the correct standard or quality. The client therefore relies on the architect to deliver the project.

Risk

There are disadvantages to this method of procurement. Although the architect is the main point of contact for both the design and construction the responsibility for risks is divided: design risks rest with the design team, risks relating to the physical product – the building itself – rest with the contractor and his sub-contractors. This means that if a building (or part of a building) fails, for example a roof leaks, the failure may be due to a design defect such as a faulty design of the guttering, or a construction defect – the guttering was not built in accordance with the drawings or the material is faulty. It will be costly to find the reason for the failure, to allocate responsibility and then to remedy it – during which time the building will continue to leak.

However, as well as design and physical building risks there are risks to be managed during the procurement and construction process itself.

You will know that risk is linked to (un)certainty and the early stages of a project are characterised by many unknown factors – problems with access outside your control, soil conditions, drainage and services crossing the site, unknown sewers or even underground streams, contamination on 'brownfield' sites as well as the risks associated with asbestos and so on.

The greatest risks, though, generally occur during the construction process itself. These include unforeseeable events such as exceptional weather but also changes made by the client to the scope of the works or the design itself. Remember that the entire design should in theory be complete at the time the contractor bids for the construction work. Because design is complex there may be insufficient information to complete the construction or inconsistencies caused by lack of co-ordination of design information resulting in requests for further information by the contractor to the architect. There may also have been pressure on time at the production information stage or the design may not have been fully developed leading to design changes *after* the contractor has been selected and appointed to carry out the works. These can all lead to the risk of increases in construction cost and the time for completion.

Summary

General contracting has to strike a balance between the value of the architect as the single point of contact during the whole process and the inherent weakness of a linear model that presupposes that the design is complete before construction commences. Also there are inherent risks to the client in a system that does not accommodate uncertainty and change easily. When events occur the risks tend to be borne by the client. As a professional the architect has to act impartially in deciding claims even though s/he is employed by the client and the claim event may be due to incomplete or late design information. Lastly the split responsibility for design and construction risks makes it difficult to point the finger and apportion blame for failure.

There are also a number of other problems with general contracting. The linear nature of design and construction means that it takes considerable time to deliver an entire project from inception to completion. As a designer you know that design is an iterative process and it is sometimes difficult to know when to 'freeze' the design. The model also requires the architect to make decisions without all the facts to hand, inevitably leading to assumptions that are no more than educated guesses. If these assumptions are wrong then the risk is that both construction cost and delay will increase.

Finally, because the architect as project leader manages all aspects of the design, procurement and project delivery it is difficult to scale up this process to large and complex projects and still manage the process effectively. This requires a range of professional management skills in addition to your design skills. At the point where projects become increasingly complex it makes sense to redefine the roles and responsibilities of the design and construction team and bring in other project management specialists.

DESIGN AND BUILD

 In design and build procurement there is a single point of responsibility. The other key aim is to reduce the cost and programming uncertainties experienced with general contracting. It has been suggested that this is the most logical way to procure a building and that given a 'clean slate' all buildings would be delivered this way.[114] When we buy a car, for example, we accept that it has been designed by its producer who takes responsibility for the design and manufacturing process and guarantees that it is fit for its purpose. The specification, performance, price and delivery time are agreed – generally. It is also rare to celebrate the achievements of the individual manufacturing designer – Jonathan Ive at Apple, James Dyson of eponymous vacuum cleaner fame and Kenneth Grange – the 'godfather' of English industrial design responsible for many familiar consumer objects are exceptions. It is even rarer to celebrate the production team of engineers that makes the product and delivers it to the supplier and end-user.

Design and build procurement pre-dates general contracting: until the idea of the 'professional architect' emerged in the nineteenth century most building and engineering projects were designed and built by the designer who also took on the management responsibilities of the contractor for delivering the project. The architect or engineer may also have been responsible for arranging finance. In short the project risks lay with one group or individual. This had its drawbacks – especially if costs over-ran or projects experienced technical difficulties. Projects frequently stopped in mid-construction as funds ran out. Later, the two roles of designer and constructor were divided and this remained the norm for building construction projects where the designer took the lead.

The design-and-build method of procurement, placing responsibility for both functions under one roof, has re-emerged as a significant method of procurement. As Murdoch and Hughes have observed:

Design and build is a logical, clear and sensible method for procuring a wide range of buildings. As a procurement method it is a realistic and worthwhile alternative to general contracting or construction management. There is no real limit on the type or scale of project for which it can be used, but it is inadvisable to use it for high-risk projects, or for adventurous schemes. It offers a high degree of cost certainty.'[115]

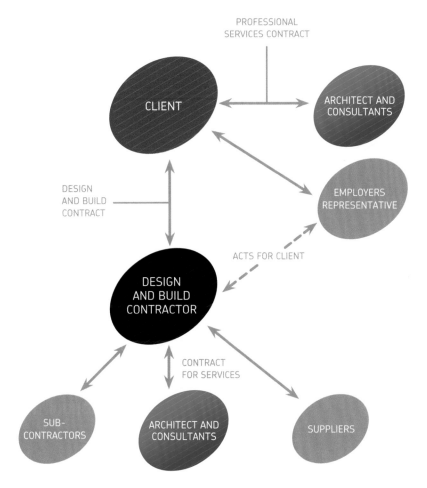

PROFESSIONAL
SERVICES CONTRACT

CLIENT

ARCHITECT AND
CONSULTANTS

DESIGN
AND BUILD
CONTRACT

EMPLOYERS
REPRESENTATIVE

ACTS FOR CLIENT

DESIGN
AND BUILD
CONTRACTOR

CONTRACT
FOR SERVICES

SUB-
CONTRACTORS

ARCHITECT AND
CONSULTANTS

SUPPLIERS

FIGURE 10.4

**Design and
Build contractual
relationships**

There is a difficulty though: there is no single definition of what we mean by 'design-and-build' other than the common idea of a single point of contact. This is because 'design-and-build' covers a huge spectrum of both design and construction. At one end there is the 'turnkey' solution where literally the client commissions a building and on completion 'turns the key' and occupies the building having had little or no involvement with the design process beyond agreeing the scope of the project. It could be as simple as a statement – the 'Employer Requirements': 'We want an office building of x square metres for y amount and I want it completed by z'. The contractor is responsible for all the design work from initial design, through detailed design to delivery and occupation. You can see that in this case the design and construction risks are taken by the contractor. The client who commissions the project has no involvement in the detail and gets a solution provided by the design-and-build team. That is not to say they will not be consulted at key stages but all the risks are managed by the contractor who remains the single point of contact.

At the other end of the spectrum the designer, working directly for the client, is continuously involved at all stages of design development: expanding the 'Employer Requirements' to include the detailed design and production information and setting out what is to be built in great detail. Only the contractual relationships change as the architect moves from being employed directly by the client to being employed by the contractor, for the construction period only, in order to create the simple 'single point' of contractual responsibility.

You can see that this procurement route, in its simplest form, is attractive to risk-averse stakeholders, typically public sector clients. It is also attractive where the risks are high and the amount of money at stake is large too. Historically, infrastructure projects such as water treatment plants, reservoirs and power stations are commissioned this way. Performance is paramount and design takes a back seat. This causes problems when design is also important or the client wants more control without accepting the transfer of risk – especially construction risk.

Project risks

It is worth remembering that project risks do not disappear – they just get moved around: transferred between clients, designers and contractors who each manage them as best they can. This is where design-and-build becomes more complex. Key stakeholders – commissioning clients – want a degree of design control although they may trade this for greater cost and programme certainty. Architects too, as major stakeholders in the project delivery, will also want to retain design control to determine how a design is developed and built to achieve the quality of the original design.

The value placed on design may also be a key factor. Traditionally design-and-build has put the emphasis on 'build' with 'design' sacrificed for cost and programme certainty. It should also be recognised that many buildings – typically industrial buildings, have almost always been procured this way: where cost and function, and not design are at a premium. The projects are also of relatively low perceived environmental value and are able to re-use or adapt tried-and-tested basic designs. Due to their location, edge-of-town industrial estates and business parks, for example – they do not attract much design interest. As a typology they are 'Cinderellas' – attracting some interest from critics and academics interested in the 'architecture of the everyday'.

Managing design risk

Over the last thirty years major private and public sector clients and other stakeholders such as funders and governments, concerned about the misallocation and mismanagement of design and construction risk in traditionally procured projects have increasingly sought to reduce these risks while retaining design quality. Architects – have also sought to maintain design control. The solution that has evolved is, in effect, a hybrid, leaving the design development and detailed design activities of the early RIBA Work Stages with the design team and transferring this risk to the contractor for the later RIBA Work Stages.

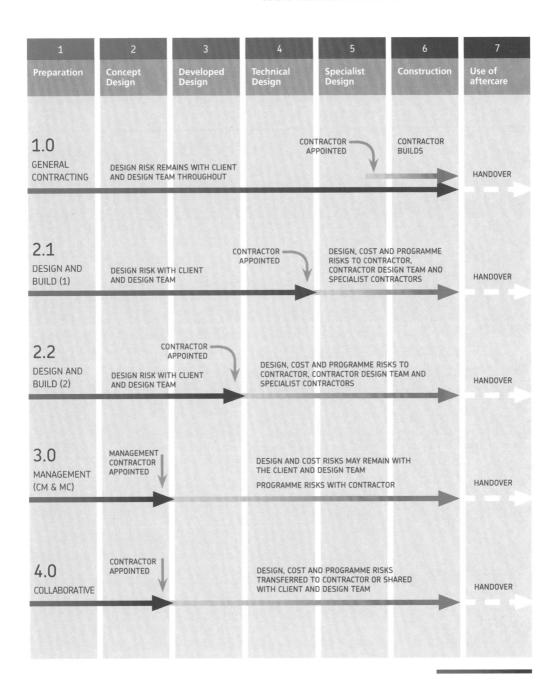

1	2	3	4	5	6	7
Preparation	Concept Design	Developed Design	Technical Design	Specialist Design	Construction	Use of aftercare

1.0

GENERAL CONTRACTING

DESIGN RISK REMAINS WITH CLIENT AND DESIGN TEAM THROUGHOUT

CONTRACTOR APPOINTED

CONTRACTOR BUILDS

HANDOVER

2.1

DESIGN AND BUILD (1)

DESIGN RISK WITH CLIENT AND DESIGN TEAM

CONTRACTOR APPOINTED

DESIGN, COST AND PROGRAMME RISKS TO CONTRACTOR, CONTRACTOR DESIGN TEAM AND SPECIALIST CONTRACTORS

HANDOVER

2.2

DESIGN AND BUILD (2)

CONTRACTOR APPOINTED

DESIGN RISK WITH CLIENT AND DESIGN TEAM

DESIGN, COST AND PROGRAMME RISKS TO CONTRACTOR, CONTRACTOR DESIGN TEAM AND SPECIALIST CONTRACTORS

HANDOVER

3.0

MANAGEMENT (CM & MC)

MANAGEMENT CONTRACTOR APPOINTED

DESIGN AND COST RISKS MAY REMAIN WITH THE CLIENT AND DESIGN TEAM

PROGRAMME RISKS WITH CONTRACTOR

HANDOVER

4.0

COLLABORATIVE

CONTRACTOR APPOINTED

DESIGN, COST AND PROGRAMME RISKS TRANSFERRED TO CONTRACTOR OR SHARED WITH CLIENT AND DESIGN TEAM

HANDOVER

FIGURE 10.5

Design and construction risk in different procurement routes set against the RIBA Plan of Work (2013)

The point at which the contractor takes responsibility for design risk can be early in the design process or very late. The design quality can also be retained by using the same design team for the construction phase but transferring the design risk to the contractor. This process is called 'novation'. In effect a 'new' contract is created where the design team (or possibly only the architect) previously employed by the Client, is employed by the contractor. Alternatively, the contractor may bring in his own design team to develop the detailed design and production information. This is called 'Consultant Switch'.

'Novation' or 'Consultant Switch'?

The decision to 'novate' the existing design team (or architect) to continue the design development, or switch consultants, will depend on a number of factors. Novating the design team brings the benefit of design continuity, a good knowledge of the project and its difficulties along with a good understanding of the design philosophy that governed the key design decision-making. However the original relationship with the commissioning client ceases. The contractor is now the architect's new client and may have a different view on design development – especially as they have now taken on the design risk.

There may also be problems associated with the design team now working on the 'supply' side being asked to do things that clash with objectives set out when they worked for the client – on the 'demand' side.

For these reasons the contractor may prefer to use his own design team. However, for the contractor there are also considerable risks in getting up to speed with someone else's designs at a critical phase in the project. It is arguable that the design risk increases rapidly during this period as the new design team tries to develop the detailed design while now working within the construction programme agreed by the contractor. Designs may be compromised purely to meet the constraints of the programme.

Construction Risk

The key reason for selecting the design-and-build procurement method is to transfer design risk to the contractor. However the contractor also takes all the construction risk. This includes design development and programming risks. Although, in theory, these risks are the same whichever method is selected, the architect no longer has a role in the construction phase other than to co-ordinate information in accordance with the contractor's programme. S/he no longer acts as the client's 'eyes-and-ears' beyond the early design phase of the project, executes client changes or checks that the construction is of the agreed quality. The architect no longer has to deal with contractor claims for additional work or time. Any deficiencies in the design due to incomplete design development or the late issuing of design information are now the responsibility of the contractor who manages these risks in order to achieve the agreed contract sum and completion date. The client therefore runs the risk that the design standards will be compromised to achieve a fixed contract value and completion date and to increase the profit margin required by the contractor.

The opportunities for change are also very limited. The client may order or the contractor may offer a change. This removes one of the major drawbacks of

General Contracting: design development after the contractor has started on site leading to claims by the contractor for changes to the project.

Stakeholder (Client) risk

One of the key benefits of the General Contracting method is that the client retains the architect to act as his or her agent ('eyes and ears') and delegates the quality control of the construction work to the architect. In design-and-build this function is no longer performed by the architect – even when they are novated to the contractor. The client therefore has to carry out this risk management function or find someone else who has the expertise to manage these risks without necessarily being a designer. Because cost risk as well as programme is usually critical to the project's success this role is normally carried out by a specialist cost consultant.

Summary

Design-and-build has evolved into a mature procurement method that provides a single point of project responsibility and effectively manages both design and construction risks. The 'wrinkly tin sheds' that used to be associated with design-and-build still exist but the method can also result in well-designed sophisticated buildings. The main attraction to professional but risk-averse stakeholders, from developers to government organisations, (other than the single point of responsibility) is that the design development risk is transferred to the contractor thus reducing both cost and programming risk. The trade-off may be a marginal reduction in design quality but it may also achieve benefits through closer working by designers and contractors – a theme developed in the following analysis of management and collaborative procurement.

MANAGEMENT

In the overview of the construction industry it was said that all contractors employ other contractors for some or all of the construction work. Because the construction industry is cyclical – depending heavily on economic trends in the wider economy – it is uneconomic to employ all the necessary trades continuously. The nature of construction is that it is project-based and each project is unique calling on different skills and resources. Construction has become more complex and international and now relies heavily on specialist contractors for items with a design element. These companies are often major international and commercial enterprises with an annual turnover larger than the construction company they are working for. For example Kone, the lift and escalator company is a global company with products that are used on many major projects worldwide. They employ 35,000 staff in over fifty counties. In 2011 their net sales were Eur 5.2 billion.[116] Wilmott Dixon, the major UK contractor, has a turnover of £900 million in 2010 and 2,600 employees.[117] Kone dwarfs Wilmott Dixon in terms of size, annual turnover and its international presence.

Over the last thirty years contractors have increasingly relied on these core specialist sub-contractors. Their commercial and project success depends upon their ability to negotiate a competitive price for each sub-contractor and to plan and co-ordinate their different tasks within the construction programme. In effect these specialists are also major stakeholders in the project too. This relationship between contractor and package contractor is reflected in and made explicit in the management procurement route. (See diagram below)

A key characteristic of the management procurement route is that the client will appoint the management contractor at an early stage in the project. The early involvement of both the contractor and possibly the specialists with the design team at the early stages of the project can have many advantages. It allows the design team to work with the specialists and exchange information before the detailed design work and production information stages are reached. It also encourages the contractor and architect to consider buildability as well as design and to evaluate different cost and design options.

There are two ways to appoint these specialists: a) as sub-contractors to the management contractor – this is called 'Management Contracting'; and b) directly by the client – this is called 'Construction Management'.

Management Contracting

In this method the management contractor is employed directly by the client as a specialist alongside the design team early on in the project – possibly at the outline design stage. He then takes responsibility for the works contractors who deliver their specialist work packages, which represents all the contract works, under his direction. It is therefore similar in many respects to General Contracting.

FIGURE 10.6

Management Contracting contractual relationships (note similarity with General Contracting)

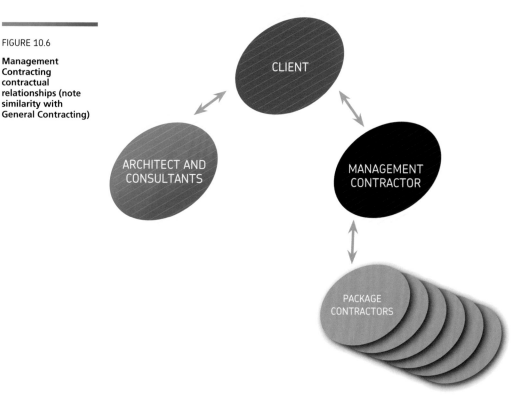

The management contractor relies on the specialist package contractors for one hundred per cent of the works. However the management of the procurement process is very different. General Contracting is linear and sequential – the design team completes its design which is in theory then fixed; the contractor bids for the work and if successful builds the project. Any design changes 'downstream' create significant extra costs and will possibly affect the construction programme. In management contracting the process is iterative: the design team can work with the construction team and package contractors at the early stages of the design to find the most cost-effective solutions. The management contractor can also negotiate with the package contractors early on before the programme is fixed. This has the advantage of being more competitive and allows the successful works contractor to programme manufacturing in advance.

Management contracting aims to retain the advantages and features of General Contracting with an independent design team. The early appointment of the main contractor and his involvement in the design development allows valuable input into the programming and cost implications of design decisions and should help to align the project budget with the actual costs retaining design quality. The method can also accommodate greater complexity and levels of change to suit alterations in the brief, design and construction.

Risk

Because the contractor is appointed early on in the process there is an opportunity to assess and manage the design and construction risks before they have an impact on the construction phase itself. Changes in requirements can be discovered early and feed back into the design development process. The main risks remain with the employer but are reduced as a result of the contractor's early appointment. The aim is to manage and reduce risks from the project's earliest stages rather than rely on the risk transfer from employer to contractor.

This works best when the contractor is truly considered another member of the project team working closely with the designers and cost consultants. However there is a tension too: the contractor still takes the same risk as a conventional main contractor under general contracting for the package contractors, programme delays and defective work.

Summary

Management contracting was intended to solve some of the problems of risk transfer in general contracting and design and build. In theory, it is therefore suited to complex, large projects with a high degree of risk. The method allows the employer to draw upon the experience and knowledge of the contractor as an expert in the construction phase who – through a non-adversarial dialogue – can highlight risks at an early stage in the project life cycle. However the relationships within the project team during the construction phase of the project are very similar to general contracting and the roles of the design team are also relatively conventional. Because the projects tend to be complex and have more risks the contractor will either price that risk or try and pass it on to the package contractors. As such it retains many of the disadvantages of general contracting.

Construction Management

FIGURE 10.7

**Construction
Management
contractual
relationships**

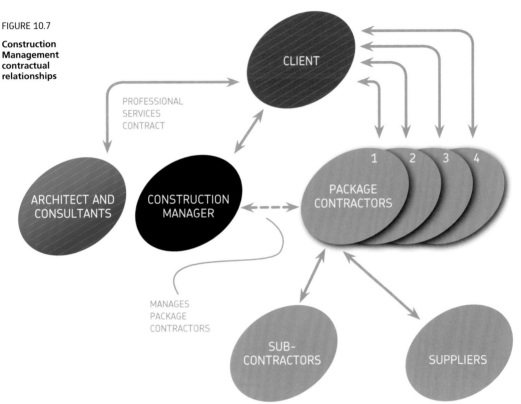

Construction management (CM) is a procurement method that fully recognises the contractor as a professional manager of the construction process in both the project team and the construction contract. The construction manager is employed directly by the client and controls all aspects of construction risk. As a professional member of the project team the construction manager can be appointed at the same time as the rest of the project team and plays the same consultancy role throughout the project. The employer, not the contractor, employs the package contractors direct. Therefore there is no contractual link between contractor and package contractors. This contrasts with management contracting where, following the contractor's early involvement, his role during the construction phase becomes similar to a conventional contractual relationship.

CM is suitable for very large complex projects with a high degree of both design and construction risk. The structure allows for early involvement of specialist package contractors during the design phase and recognises the expertise of major package contractors such as Kone (in the example above). It fully recognises that the success of a project is dependent on the performance of the package contractors and their close involvement in the specification, design and programming of the project. It is also transparent and allows the employer to seek the best value by negotiating direct with the package contractors.

Risk

The main disadvantage is that because the employer contracts direct with the package contractors all the design and construction risks sit with the employer. Therefore it is only suitable for experienced and professional project clients who are used to commissioning projects and have a good understanding of construction projects. These include 'serial commissioners' such as large development companies but also include large universities and employers used to managing capital projects such as energy, infrastructure and pharmaceutical companies.

To manage the risks effectively the employer relies upon an experienced project team who have the resources and expertise to work in this more fluid contracting environment. As there is no single main contractor to give a contract price for the works the client will increasingly rely upon the cost consultants to prepare, monitor and update a Cost Plan. There is also a significant co-ordination exercise at both the planning and construction phases to reduce the risk of time over-runs.

The upside of this more sophisticated approach to procurement is that risks can be identified and managed at the earliest possible moment in the process. Design risk can be identified and managed though early involvement of the specialist package contractors in, for example cladding or advanced sustainable services design decision-making. The design risk can be transferred to the package contractors if appropriate. This early dialogue can also flag up construction co-ordination problems and lead to efficiencies in the project both in terms of the product and the process. These efficiencies can lead to cost savings which are then passed on directly to the client as direct employer.

Summary

Construction management is suited to large, complex projects commissioned by experienced, knowledgeable clients who have a sophisticated approach to risk management. The construction manager is appointed as a member of the project team as a professional expert in construction. Because the package contractors are appointed directly by the employer all the team contribute to the management of risk: both design and construction. Because there is no overall contract the method relies upon a robust cost plan and schedule which must be continuously monitored. The transparency of the process and the early appointment of specialist package contractors who may also be largely responsible for specialist design elements of the project can lead to better risk management, greater efficiency in working practices and increased value.

COLLABORATIVE PROCUREMENT

Throughout this chapter one of the key themes has been 'risk transfer' – when and how can the client offload as much risk as possible to the contractor?' You will have seen that Construction Management recognises the value of closely managing the design and construction through early involvement of works contractors. A further step in risk management is to propose that risk should rest with whoever can manage it most effectively. This could be one or more direct stakeholders in the project: client, design team, construction manager and works contractors. This more collaborative approach involves risk allocation and possibly risk sharing and risk retention by the client as well as risk transfer to the construction manager or package contractors.

The benefits of this approach are that it encourages a team approach to problem solving rather than an adversarial allocation of responsibility. It also recognises the strengths of stakeholders and potentially encourages them to take responsibility for solving problems where in the past they may have sought to pass them on. It encourages a more open approach to pricing the work and to establishing long-term partnerships. This avoids the 'learning curve' that all stakeholders refer to at the project start-up stage and has the potential for shortening lead-times. It is also suited to 'professional clients' who understand project processes and are prepared to take on project risks themselves in return for greater transparency and efficiency.

An example might be a new laboratory building on a university campus. The client is best placed to understand the problems of health and safety on a fully functional and occupied campus. They may already work closely with specialist equipment suppliers who also maintain the existing plant. It is a short step for the client to share the health and safety risks with the construction manager and to take responsibility for the specialist package contractor. You will see that this only makes sense if the client is able to effectively manage these risks. If not, then they should be transferred. However a more collaborative approach will open up a dialogue early on that may lead to efficiencies and enhanced site safety.

This approach is not for the faint-hearted. The 'warts-and-all' approach to risk sharing has the potential for more creative problem-solving but requires stakeholders to leave many of their attitudes to risk transfer behind and take a chance on a mutual approach to project problem-solving. A collaborative approach requires stakeholders to adopt a change in attitude to make the method work. In the Constructing Excellence contract, for example, 'Working Together' is a key objective governed by an 'Overriding Principle' that guides key stakeholders:

'…to work together with each other and with all other Project Participants in a co-operative and collaborative manner in good faith and in the spirit of mutual trust and respect.'[118]

CONSTRUCTION CONTRACTS

It is beyond the scope of this chapter to discuss construction contracts in detail. However it is worth building on some of the principles of contract discussed elsewhere in the book and to place them in the context of procurement. A contract is a legally-binding agreement that sets out responsibilities, commits the parties to performance of the terms of the contract and allocates risks. It is a key principle of construction contracts that where risk is accepted this is reflected in a cost premium: as more risk is transferred to the contractor the price goes up. Therefore in a design and build contract the contractor will charge a premium for accepting the design risk. In a more collaborative approach as the risks can be shared or transferred to the employer so the premium that the contractor will normally charge can also be shared, thus reducing the cost to the employer.

Over time different contracts have been developed for each procurement method and typically the prominent stakeholders in projects have met to agree how to allocate risks to suit each procurement method. The organisation that acts as a forum for discussion and agreement on construction contracts in the UK is called the JCT: the Joint Contracts Tribunal.[119] It brings together representatives from the clients, consultants and contractors to agree common terms for different types of construction contract. This has led to a 'family' of standard contracts which the JCT then publishes for use by major stakeholders. Examples include the following:

General Contracting: JCT Standard Building Contract (SBC)

Design and Build: JCT Design and Build Contract (DB)

Management: JCT Construction Management Trade Contract (CM/TC)

JCT Management Building Contract (MC)

Collaborative: JCT Constructing Excellence Contract (CE)

There are other stakeholder organisations that also produce construction contracts. These include the Institute of Civil Engineers (ICE) in the UK who publish the New Engineering Contract (NEC)[120] – a collaborative management contract which was written in direct response to the issues raised in the Latham Report. In recognition of the global nature of construction there is also a family of standard international contracts published by FIDIC[121] which are used on major infrastructure projects and by major international stakeholders such as the World Bank who fund many public sector projects.

In each of these contracts the roles and responsibilities of stakeholders are different depending on the attitude to design and construction risk. Scale is also important. For example, the architect retains a prominent role in the General Contracting JCT Standard forms but roles and responsibilities vary significantly in the design and build and management forms of contract.

Because construction contracts allocate responsibilities that are legally enforceable it is important that the correct form of contract is used. The architect is not expected to be an expert in construction law but you must understand the architect's duties under these contracts and it is important that you take further advice and consult the various specialist construction contracts reference points such as the JCT or NEC before selecting one for a project.

COMPETITIVE TENDERING

This section covers the way that a contractor is selected to carry out the contract works. This traditionally involves inviting different contractors to bid for the construction phase of the project. The main objectives are to obtain the best value for the project from contractors who are best able to carry out the works in the time and to the standard required by key stakeholders. Here 'value' means quality, time *and* price. It is important to ensure that the tendering process takes all these factors into account rather than price alone. There are a number of ways to tender for the complete project or parts of the work depending on the procurement method selected. The key principles that govern tendering remain the same: fairness and transparency with the objective of gaining the best value. A useful source of 'best practice' on tendering is the 'NBS Guide to Tendering: for construction projects.'[122]

Single Stage and Two Stage selective tendering

Because tendering is so important and provides the transition between design and construction major stakeholders have taken the same approach as they have to construction contracts and created a forum to agree standard methods for tendering. As a result a number of detailed standard procedures are published that are followed by stakeholders.[123] This section outlines the basic single stage selective tendering procedure for general contracting. From this foundation it then discusses two-stage tendering and lastly how European legislation has affected tendering for most public sector projects. Although general contracting is used as an example, the methods apply equally to other procurement methods.

Single stage selective tendering

Single stage selective tendering follows the sequential project process outlined in General Contracting. The design team, in theory, produces a comprehensive set of documents which includes the detailed design, a detailed specification defining quality standards and scope, any specific site conditions and hazards and a health and safety plan. The documents may also include schedules of standard work activities or rates to, for example, excavate ground or lay bricks.

Contractors may be invited to go through a pre-tender selection exercise that assesses their suitability to carry out the works. From this exercise a shortlist of around six contractors is created. Identical documentation is then sent to the contractors on the shortlist inviting them to submit a 'fixed price' tender for the works. (Sometimes the documentation will state 'provisional sums' for work that has not been fully designed or specified, for example sanitary fittings or finishes that the client has not made a decision on. Costs for these items will be priced by the contractor during the project works. You will see that there is a danger of specifying too many 'provisional sums' as the contract price will not be fixed.)

Tenders are then returned in a sealed envelope and these are opened at a meeting attended by the client and key members of the design team. Tenders are then scrutinised for errors (there are standard methods for contractor to address errors found in the tenders). One key principle is that contractors only submit a tender for what has been asked for. Therefore tenderers should not qualify their tender by saying, for example, they have looked at the specifications but their price is based on the client accepting a different specification or supplier for, say, the lift installation or the type of air-conditioning. The reason behind this is that it is no longer possible to make a fair assessment because the tenderer has priced the works differently. Following acceptance of the tender the contractor will be appointed to construct the works within a set period of time stated in the tender documents or agreed as part of the tender submission. It is important to note that the construction contract will include the priced tender documents. In other words the tender documents become part of the contract and bind the employer and contractor to what was specified and act as a benchmark for the project. Any changes made by the client or the design team will potentially be a variation to the contract and run the risk of increasing the cost or construction programme.

Two-stage selective tendering

Two-stage selective tendering has evolved to acknowledge that not all construction information is available at the point contractors are invited to tender for the works. It also recognises that there are benefits in appointing the contractor before the design is finalised. The shortlisting takes place in the same way as for single stage selective tendering but the information that is priced is very different. The objective is that tenderers will be selected competitively on certain known parts of the work at the first stage. The selected main contractor will then contribute to the second stage tendering for the different works packages and may suggest alternative ways for the packages to be defined and will agree the timescale for tendering to meet the construction programme. The works contractors will then be employed directly by the main contractor who will manage them.

Although the procedures for two-stage tendering are standardised unfortunately the point at which tenders are invited in design development is not. Therefore two-stage tendering may include a variety of information from a brief outline of the works to very detailed production information for some works packages but limited information for others. The risks escalate in inverse proportion to the design and production information available. The client and the design team have to weigh up the advantages of an early main contractor appointment over the risks associated with the uncertainty caused by the lack of fixed and comprehensive information. The conventional argument is that the tendering process is still active and in some instances the contractor may not be appointed for the second stage. (In practice this only increases the construction risk and negates the value of an early appointment.) The key difference is that as the contractor is now on site and has a programme works package tenderers will have to meet the programme – often with shortened lead-in times. This in effect limits the tender choices at the second stage and may not be as competitive as singe stage selective tendering. It also transfers a significant amount of programme risk to the design team: a failure to deliver the tender information package on time will adversely affect the construction programme. Two-stage selective tendering is not therefore an easy option and has to be managed carefully by the design team. Delays can lead to claims and a more adversarial approach to the construction phase.

Public sector tendering and the European Union

For many years the UK construction industry has agreed its own standard procedures for tendering and these have been broadly acceptable to all stakeholders in the private and public sectors. However, one of the main objectives of the European Union has been to open up markets to all member states. It is natural that clients, especially public sector clients spending taxpayers money will tend to favour local contractors. The Single Market targeted tendering procedures as a barrier to trade between member states and the result is a transparent tendering procedure. All public sector projects above a certain contract value, including consultancy contracts for architectural services, are advertised in the OJEU (the Official Journal of the European Union – also called OJEC). The EU sets out a transparent and fair selection process which enables stakeholders to shortlist suitable companies to tender for consultancy and construction work contracts.

Unlike the NBS Guide which sets out voluntary 'best practice' adopted by the construction industry in the UK, the OJEU procedures are mandatory and a failure to follow the procedures fairly can result in prosecution. The statutory rules are set out in the Public Contracts Regulations 2006 implement into UK law an EU directive which was adopted in 2004.[124] The rules are complex and stakeholders have been prosecuted in the UK and across the EU for failing to follow the procedures.

COST AND VALUE: ESTIMATION AND UNCERTAINTY

As architects and designers you will not be expected to be experts in cost estimation. As you already know most projects – even small ones – are potentially complex. On larger projects stakeholders will rely on expert cost consultants. (They will also be very useful on smaller projects too – if clients are prepared to pay for their services.) However cost consultants will rely upon the rest of the design team for as much information as possible. They cannot create costs from thin air. Construction projects are characterised by fluidity and uncertainty. As cost is an expression of both value and risk the fuzzier the boundaries of a project, the greater the uncertainty and risk of the costs being either over- or under-estimated. Cost consultants use a number of methods to estimate cost under uncertainty. In particular they will use historic data for similar projects. This can translate into a cost per square metre with additional sums of money to value additional perceived risks. However the size of a project is often not known at its earliest stages when you are testing its viability.

They can also use an elemental method where the project is broken down into smaller elements for which industry-standard data is available. This depends upon an unambiguous interpretation of the scope and materials. For specialist items such as cladding or services they can tap into the expertise of specialist works contractors and their databases of previous projects.

The key role of the architect at this early stage is to explain to the team, and the cost consultant in particular, the key features that bring value to the project and those issues that may affect cost. In particular, where a novel or untested element is being used then this should be emphasised so that reliance on historic data from apparently similar designs can be modified. (It may also be necessary to explain the value that comes with the design too.) The architect's response to requests for information should be as comprehensive as possible and you should welcome the opportunity to engage with the cost planning process as part of the development of the project brief. Of course, supplying good information is particularly difficult in the early stages when the design is still highly fluid. It is a truism that design changes made early in the project cost very little – if nothing other than a change to the budget. Late changes to the design, especially at the construction phase, are significantly more expensive and should be resisted. You should think of cost not only as a product of your design decision-making but as a factor or constraint of the brief similar to a physical constraint.

One method of capturing attitudes to value and costs is to hold value management workshops at critical phases of the project. This gives all stakeholders, including designers and clients, the chance to share what they value in the project and agree what costs are acceptable. This may result in negotiation and possibly changes to the brief – or the design. If a consensus is achieved around a comprehensive cost plan that recognises both the value of the design as well as the costs there is a greater likelihood that the project will meet both the cost plan and the brief.

REFLECTIONS ON PROCUREMENT

Procurement is the process by which your design is transformed into a built form. This transformation can only take place by engaging with other key stakeholders – in particular clients, other designers, cost consultants, health and safety experts, specialist contractors and experts in the construction process itself.

Construction projects are fluid and with that fluidity comes uncertainty and risk. The role of the architect changes according to the size and complexity of the project from the single point of contact and risk manager on small projects to that of design specialist on larger projects where the risks are too great to be managed by a single individual and requires a team approach.

The different methods of procurement have evolved to manage and transfer risk amongst key stakeholders – the client and contractor in particular. However risk does not disappear until the project is completed and then there are residual performance risks which exist during the lifetime of the building. Contractors, in turn, charge a premium for accepting design and/or construction risks.

The building contract is the way that the burdens and risks are agreed between the employer and the contractor and set down in writing. It follows that the contract must reflect these accurately and that the architect must be aware of the significance of each type of contract.

The procurement loop is completed by the tendering process which has the objective of achieving best value in the marketplace.

FURTHER READING

Murdoch J & Hughes W 'Construction Contracts' (4th ed.) Routledge 2007

www.jctcontracts.com

Finch R 'NBS Guide to Tendering: for construction projects' RIBA Publishing 2011

11
THE FUTURE

The main aim of this book is to establish professional studies as a clear subject area within the architectural curriculum. It would be wrong to say that the book integrates the subject topics within your design work. However, it shows that architecture – as a creative and socially-motivated endeavour – requires a knowledge and understanding of the professional and legal context within which it operates. It also requires an understanding of the complex processes that take your work beyond the design studio and place it firmly within the wider context of the construction industry.

One of the themes of the book is the role of the architect as a co-professional. The **professional architect** is a specialist, with specialist skills and knowledge. In common with other professions, architects historically enjoyed a particular status in society with barriers to entry to the profession, mutual recognition from other professionals (and, more importantly, professional organisations) and ethical standards that aimed to demonstrate that they worked in the public interest rather than self-interest. It was, in effect, a trade-off: status for public interest. Increasingly this trade-off has been questioned as some professionals have been shown to work in their own interest first and to use their specialist knowledge to manipulate professional and legal rules. The idea that only fellow professionals could police their own has been eroded and now professional standards are generally monitored by independent bodies set up for that purpose. We live in a more egalitarian and consumer-led society where the service provided by professionals generally is questioned more readily.

Despite the erosion of professional status, there is now an even greater role for professionals as specialists: highly trained and competent to deal with complexity on behalf of their clients. The complexity of the construction industry has led to the emergence of new professionals in design, management and construction. The traditional role of the architect taking a client through all the activities needed to design, procure and deliver a building still exists but is now itself a specialism sitting with other specialist architectural roles. It could also be argued that architecture remains a profession that truly works for the public and in their interest: setting stylistic battles aside, the public experience the architectural profession's talents and skills in a way that extends far beyond the reach of other professions. Also, the work is always on view – open to scrutiny and public debate in some instances.

The public sees a product: the building, urban design or landscape but not the complex **design management** processes that are required to make the product happen. In one sense this is no different from complex civil engineering projects such as bridges and tunnels or engineering design and production management that contribute to the design and manufacture of mobile devices and apps. They all require an organisation and specialist processes to support the design process. On the other hand, construction as a process appears more fluid and variable, with multiple unknowns caused by the fact that each project is unique: with a specific site and its own design that responds to the site. Design is a highly iterative process that has to fit into a more linear model of the construction delivery. This inevitably causes tensions in the total construction process and raises many challenges for effective design management. Within this design process there are multiple stakeholders who have to be managed, their voices heard, and requirements met – all within finite budgets and timescales. Design management therefore requires its own set of knowledge and skills.

The complex relationships between stakeholders and their attitudes to risk and uncertainty have led to different methods of procurement. Buildings are also highly complex, requiring design specialisms that go beyond the architect's design knowledge and competence. Some of these skills rest with specialist contractors. These include specialist elements such as cladding, lifts or the environmental systems on which buildings depend. These separate designs and construction skills have to be integrated and managed within the total delivery. New professionals such as construction managers have achieved the competence to deliver this service as part of the design and delivery team. Architects, as designers and design managers, need to understand the wider context of delivery in order to contribute to the decision making process.

The complex design and procurement processes sit within the **English legal system**. The Common Law is at the heart of the system: a set of procedures and enforceable rules that have developed over centuries which adapt to changing attitudes and values in society. These enforceable rules affect the way we establish legal relationships through our contracts with key stakeholders: from clients to construction contractors. The contracts and agreements which we enter in to are essentially private ('private law' as it is sometimes called) and set out responsibilities and expectations for performance and payment. In addition the common law sets rights and our wider duties to society to ensure that our rights are respected. Even if we do not have a contract in place we owe wider duties to respect other people's land and their right to privacy, for example.

In addition to the common law, Parliament steps in to modify our rights for the benefit of society or to use the legal system to enforce political initiatives though Acts and the regulations that flow from them. For example, the Human Rights Act 1998 was the result of a political initiative that introduced widely accepted international human rights into English law and therefore added to and modified the historic rights established by common law. The town planning system can be interpreted as an example of Parliament stepping in and altering our common law rights over the land and buildings we occupy and own. This set of complex constraints now controls the way we use land and develop the built environment. Architects need to be aware of and work within this set of constraints. Our participation in Europe and world trade has also led to the integration of European law and international laws and protocols. Construction is international and relies on international capital and trade and therefore has to comply with these other legal systems where they are relevant.

Building development and economic growth were the original drivers for the post-World War Two society but the planning system has more recently been adapted to recognise the need to preserve our **heritage**. This protection of buildings, the urban grain and landscapes has led to another layer of complexity which architects must understand and navigate through. More recently, local and global **sustainability** has become increasingly important to the decisions that we take about land and energy use and the waste and pollution that we produce. The built environment is a major consumer of land and energy and architects have a role to be proactive in trying to balance economic growth, the responsible use and preservation of finite resources and minimising waste, not only of CO_2 but other waste products.

THE FUTURE

Futurology is a dangerous pastime. The current model of the architect as a construction professional, including the pattern of architectural education – the foundation of your knowledge and skills as a design professional and the RIBA Outline Plan of Work – the basic model of design management, was developed in the 1960s. If you are reading this book at the beginning of your career in architecture then you might still be in practice in the 2060s, by which time the model will be about 100 years old. It might be argued that it has shown its value by adapting to change and will continue to stand the test of time. However, in a society where the notion of the profession reflecting a set of values propped up by mutual recognition of other professionals is under such intense scrutiny, it is possible that the professions will disappear in all but name and stand for particular specialisms instead.

Fortunately, regardless of the changing face of the professions, society's unending need to create and modify the environment will mean that the particular skills of the architect will continue to be required. How architects of the future deploy their design talents will depend on their understanding of the complex legal framework within which they operate and their ability to manage the design process adapting to the changing procurement environment.

In everyday architectural practice, you will require a deep knowledge of current law, regulations and procedures, all of which are inextricably part of being a successful architect. This book forms the foundation you need for understanding this rich, complex and fascinating professional context. Digesting its messages will prepare you for finding out more, much more, and hopefully enjoying the challenge along the way.

REFERENCES

1   McKinlay (1973) quoted in Macdonald KM 'The Sociology of the Professions' 1995 Sage. Macdonald is a good point of reference for an overview of views and theories about the professions.

2   'The moral principles or a system of a particular leader or school of thought; the moral principles by which any particular person is guided; the rules of conduct recognised by a particular profession or area of human life.' [Shorter Oxford English Dictionary 5ᵗʰ Edition 2002.]

3   An oath derived from Hippocrates (c.460–c.370 BC) the Greek physician credited with writing it. He is considered to be the father of medicine. This quotation from part of a modern version gives a flavour of the commitment to the public interest: 'I will remember that I remain a member of society, with special obligations to all my fellow human beings, those sound of mind and body as well as the infirm.' [Louis Lasagna, Dean of the School of Medicine at Tufts University, 1964 [www.mediterms.com accessed 24.04.2010].]

4   Alain de Botton: 'Tate Britain: a symbol of Britain as it would like to be' *The Times* 24ᵗʰ April 2010. De Botton perhaps gets carried away – but his prose, in turn, shows the unique power of architecture at its best to illuminate and inspire: 'The purpose of the Tate is to evoke valuable states of mind which we theoretically approve of but forget in the run of daily life.'

5   For a further discussion see also Brookhouse S. 'Part 3 Handbook' p. 4.

6   RIBA mission statement, July 2005 (available at www.architecture.com).

7   The Privy Council is one of the oldest parts of Government. It deals with items of government business which, for historical reasons fall outside ministerial departments. Much of its work is concerned with the affairs of the chartered professional bodies, the 900 or so institutions, charities and companies who are incorporated by Royal Charter. 'The Privy Council also has an important part to play in respect of certain statutory bodies covering a number of professions and in the world of higherappendi education'. For more information see www.privy-council.org.uk.

8   RIBA Strategy 2005; www.architecture.com

9   www.arb.org.uk

10   www.arb.org.uk

11   www.arb.org.uk 'About us'.

12   Arb 'Architects Code: Standards of Conduct and Practice' September 2009.

13   That does not exclude the ARB from taking an interest in the wider issues of practice. This is expressed in some of the ARB's Standards and in the way architects can be disciplined for failing to meet the standards expected in practice. For example, an architect can be disciplined for failing to adequately supervise others in his or her office.

14   Larson MS 'The Rise of Professionalism – a sociological analysis' 1977 University of California Press.

15   Larson (quoted by MacDonald).

16   After Macdonald 'a working theory of the professions'.

17   The SEC investigation in the USA found that Worldcom used fraudulent accounting methods to mask its declining earnings. The fraud was achieved by under-reporting costs and inflating revenues. The fraud was identified by an internal team working in secret. It was estimated that the company had inflated its total assets by $11bn. Its accountants, Arthur Andersen, one of the largest accountancy consultancies in the world withdrew its audit opinion – and subsequently collapsed with its professional reputation in tatters.

18   'The Architect and His Office' RIBA 1962. The title itself is an indication of another aspect of professionalism that has not been addressed in this chapter – gender – and its role in the professions.

19   'The Strategic Study of the Profession' RIBA 1992. The Study was much more systematic than previous studies and used a number of interest groups to arrive at its conclusions.

20   Stephen Hockman QC, chairman of the Bar, the professional body for barristers, speaking at their annual conference in 2006 reported in *The Times*, 6ᵗʰ November 2006. Barristers have almost unequalled protection of title and function.

21 These ideas are discussed in a paper by Davies Wand Knell J in 'The Professionals Choice' 2003 Cabe, London.

22 Quoted from an address by Norman Foster to an audience of engineers at Imperial College cited in 'Building the Connection' Imperial College 2009.

23 Richard MacCormac, past president of the RIBA commenting on the challenges of architectural practice.

24 This chapter does not discuss taxation and the various advantages and disadvantages of being self-employed or an employee. This is a vast and complex area of the law. Tax laws also change frequently to try and remove perceived advantages.

25 Partnership Act 1890 Sec.1.

26 See also www.hsbc.co.uk The well-known high street banks are a good, if dry, source of information on business planning. In researching business planning it is important to remember that professional services businesses – like architectural practices – are very different to small manufacturers or retailers.

27 One of the challenges of a new practice is what to show on your website when you have built nothing so far. If you decide to use previous projects or designs produced in another office you must get their permission – the last thing you want is a claim for copyright infringement – it does happen.

28 For further information you should refer to the RIBA Code of Professional Conduct (www.architecture.com) and the ARB Code and Standards (www.arb.org.uk). The RIBA also issues a set of Guidance notes on all aspects of practice. With reference to professional agreements ARB Standard 4.4 sets down the minimum requirements.

29 See ARB Code and Standards and RIBA Code of Conduct.

30 Ajaz Ahmed interviewed in 2012 by Edwin Smith. Source: telegraph.co.uk/finance 22nd July 2012. Ahmed set up the business with James Hilton in 1994. It was sold to WPP in July 2012 by which time it employed 1,160 staff in eight offices around the world and had a revenue of $230 million.

31 Sinclair D *Leading the Team: an Architect's Guide to Design Management* RIBA Publishing 2011.

32 Alan Roberts IID Architects.

33 Report prepared by PwC for G4S www.G4S.com

34 The RIBA Plan of Work (1963) is the most enduring product of an RIBA study of the profession in the 1960s 'The Architect and His Office' (1962).

35 You may also be familiar with the different phases of the outline RIBA Plan of Work as it forms the basis for recording your periods of professional developments in the format of the RIBA pedr (Professional Experience and Development Record www.pedr.co.uk).

36 Alan Crane CBE lecture: University of Westminster January 2012. This was one of the conclusions of a post-Latham study commissioned by the government into the construction industry. Alan Crane.

37 Rebecca de Cicco www.thenbs.com/bim/topics/articles accessed August 2012.

38 www.bimtaskgroup.org accessed August 2012.

39 Sinclair D op cit.

40 Blyth & Worthington 'Managing the Brief for Better Design' (2nd ed.) 2010 p.3.

41 John Worthington.

42 Latham M 'Constructing the Team' 1998 (SO).

43 Example based on Model Strategic Brief Blyth & Worthington p. 214.

Management as a separate discipline has grown in order to address expectations and concerns of stakeholders in the context of a complex international business environment characterised by uncertainty and risk.

44 Oxford Dictionary of Law 2003.

45 Extracts from Wakefield Court Rolls 1274: Common Law Crimes.

46 Beatson J Anson's law of contract (28th ed.) (2002) quoted by McKendrick E 'Contract Law' Oxford (2003) p.4.

47 Usually the agreement is between two parties but in certain circumstances there may not

actually be an agreement: the contract will be set out by one party only. This is called a 'unilateral contract.' They can crop up in construction contracts where a 'letter of intent' is sent to a contractor that informs the contractor that their tender is accepted and that there is an intention to enter into a contract. (For more information see Uff J 'Construction Contracts'.)

48 Jordon's case against Vodaphone in 2003 was widely reported: www.telegraph.co.uk accessed 25.10.12; www.lawdit.co.uk accessed 25.10.12. The judge, Mr Justice Langley described Jordan as a wholly unsatisfactory witness. in relation to Jordan's oral evidence that there was a number of 'blatant inaccuracies' that when these were exposed he was 'reduced to embarrassed silence' in the witness box. Jordan makes a living reporting on Formula 1 for the BBC.

49 The classic definition is as follows: 'A valuable consideration, in the sense of the law, may consist either of some right, interest, profit or benefit accruing to the one party; or some forbearance, detriment, loss, or responsibility given, suffered or undertaken by the other.' Lush J Currie v. Misa (1875) quoted in McKendrick 'Contract Law' p. 160. This is a good example of how the common law evolves and legal precedent informs future case law and commercial practice. It is also interesting that this case took place when the principles of modern contract law were being created.

50 Limitation periods. It might seem confusing that the Limitation Act provides three different time periods. The difference between 6 and 12 years is determined by whether the contract is 'simple' (6 years) or formed as a deed (12 years). The time limits apply to when a claim may be made. A longstop period of 15 years applies to claims in the law of tort discussed separately.

51 The word 'tort' comes from old French and means harm. The word 'twisted' has the same origin. It is therefore appropriate that the law of tort concerns loss or suffering, harm and physical damage.

52 'Negligence' as defined by the Oxford Dictionary of Law (2003).

53 A collateral contract is normally called a collateral warranty. A warranty is a promise in a contract. This is a complex area of construction law but the same principles of contract law apply. There is also a statute that covers third party rights: the 'Contracts (Rights of Third Parties) Act 1999'. Unusually this is a piece of legislation that you can agree to contract out of – in other words if you want it to apply then you both have to agree to it. In practice it is often excluded from contracts and instead clients rely on carefully drafted collateral contracts.

54 Oscar Wilde 'The Soul of Man under Socialism' Fortnightly Review February 1891 quoted by Buder S 'Visionaries & Planners 1990 OUP.

55 A Royal Commission was set up by Parliament to investigate and report on particular issues. They generally operated by taking evidence and produced a report with a number of recommendations. Although a good idea in principle, Royal Commissions took a considerable time to report their findings – especially on complex issues. As part of an overhaul of Parliamentary procedure their use was superseded by the 'Select Committee' system which allowed for better focus on particular issues and a faster reporting structure.

56 John Burns, quoted by Cullingworth B & Nadin V 'Town and Country Planning in the UK' (14th ed.) Routledge London 2006.

57 Quoted by Cullingworth B and Nadin V p.18.

58 Morgan K 'The English Question Regional Perspectives on a fractured Nation'. p.4 www.devolution.co.uk

59 The key report which set the agenda for the post-war Labour administration was the Beveridge Report (1942) which set out a Plan for Social Security and the foundation of the modern welfare state.

60 'Baedeker Raids' These were reprisal raids by Hitler after the bombing of Berlin, so-called because architecturally-attractive cities were chosen by referring to the well-known Baedeker tourist guides.

61 The Green Belt policy might be popular with the public but both economists and planners are concerned that the policy distorts land values and constrains development. Interestingly the 1955 policy was driven through by one government minster – Duncan Sandys – who, sensing the public mood, went against the more pragmatic, considered advice of his civil servants.

62 These figures are widely published – quoted in Sport England: 'Sport in the Green Belt' Planning Bulletin 13 (2003).

63 Prime Minister Harold Macmillan at a speech in Bedford 20th July 1957. www.bbc.co.uk/onthisday

64 The Buchanan Report had been commissioned in 1960 and Buchanan brought together a group of academics and designers to work on his radical proposals for traffic congestion. The team even included one of the founder members of Archigram. The Report was influenced by American ideas about traffic planning and creating a grid with nodes to keep traffic moving. The same grid/node principles were later applied to the final plan for Milton Keynes, the last major New Town, built in the 1970s where a strategic decision was made to design a road network for the car rather than invest in a fast neighbourhood public transport system, as originally proposed.

65 The development of Tottenham Court Road in central London – from the northern end on Euston Road to Centrepoint in the south show the effects of private developer and public transport policy working together. The piecemeal development on the West side shows a dogmatic approach to road improvement with new office buildings set back from the existing road line to accommodate massive future road improvements.

66 It is government policy to replace PPGs 'Planning Policy Guidance' with PPSs 'Planning Policy Statements, the latter representing a shift towards stating a policy position rather than providing guidance that may be re-interpreted. The most valuable source of information on current planning matters in the 'Planning Portal' www.planningportal.gov.uk

67 www.communities.gov.uk NPPF accessed 23.9.12.

68 www.communities.gov.uk Planning, building and the environment: the NPPF accessed 23.9.12.

69 Larkham P 'Preservation, Conservation and Heritage' in 'British Planning' Cullingworth B (ed.) Athlone Press 1999 p. 105. Larkham's paper gives a good critique of the growth of the 'heritage' movement.

70 SPAB website www.spab.org.uk accessed 24.5.12.

71 National Trust Act 1907 p. 4 www.nationaltrust.org.uk

72 See Cullingworth and Nadin p. 288.

73 Quoted by Cullingworth and Nadin p. 288.

74 www.english-heritage.org.uk/caring/listing accessed 24.05.12.

75 www.english-heritage.org.uk accessed 24.5.12.

76 'Traffic in Towns', 'The Buchanan Report HMSO 1963.

77 www.english-heritage.org.uk accessed 24.5.12.

78 'The Battle for Covent Garden' Seven dials trust www.sevendials.com accessed 22.5.12.

79 Obituary written by Richard Rogers 'The Guardian' 17th December 2009 www.guardian.co.uk

80 Larkham P op cit. p. 115.

81 National Trust website www.thenationaltrust.org.uk accessed 13.5.12.

82 English Heritage website www.english-heritage.org.uk accessed 13.5.12.

83 UNESCO World Heriitage List www.unesco.org. The list includes 725 cultural, 183 natural and 28 mixed 'properties' in 153 member states.

84 UNESCO World Heritage Sites Criteria for selection. www.unesco.org

85 Warsaw citation Unesco World Heritage List www.unesco.org

86 EIA Directive (85/337/EEC) ec.europa.eu/environment

87 United Nations Economic Commission for Europe www.unece.org 'The Espoo (EIA) Convention sets out the obligations of Parties to assess the environmental impact of certain activities at an early stage of planning. It also lays down the general obligation of States to notify and consult each other on all major projects under consideration that are likely to have a significant adverse environmental impact across boundaries.' Accessed 24.5.12.

88 www.europa.eu/environment/eia accessed 24.5.12. In 2009 the Directive was amended to take account of plans for carbon capture and storage.

89 'This Common Inheritance' quoted by Larkham p. 261.

90 'The Natural choice: securing the value of nature' quoted by the Environment Agency. www.environment-agency.gov.uk

91 www.wwf.org.uk accessed 23.9.12.

92 WWF www.wwf.org.uk

93 World Wildlife Fund www.wwf.org uk

94 www.wwf.org.uk Living Planet Report May 2012.

95 Blowers A 'Environmental Planning for sustainable development p. 52.

96 www.hse.gov.uk/statistics accessed 23.9.12.

97 See Michael Leapman 'Inigo: The Life of Inigo Jones: Architect of the English Renaissance' 2004 (Headline) p. 183–185.

98 A rare example of pre-Great Fire London building with jettied construction survives in Strand opposite the Law Courts in central London. The huge ornate brackets that projected beyond the building line at eaves level were often re-used as ornate brackets for door canopies. Examples can be seen in the Georgian streets off Lambs Conduit Street in central London.

99 www.planningportal.gov.uk/buildingregulations

100 The main case was Anns vs. London Borough of Merton.

101 www.planningportal.gov.uk/buildingregulations

102 Regulatory Impact Assessment November 2002 www.communities.gov.uk

103 The principles of negligence are covered in an earlier chapter: 1. Duty of care; a causal link; proximity; and physical injury or harm. In the workplace the duty of care is established by the legislation. There still has to be a causal link but this is normally determined by the facts. In the case of 'proximity' the workplace setting suggests proximity but it can be seen in the Ikea case that it applies to casual users and visitors too. The facts will establish the level of physical injury or harm.

104 *The Daily Telegraph* 4th June 2012.

105 European Directive 92/57/EEC (the implementation of minimum safety and health requirements at temporary or mobile construction sites).

106 These terms are specifically referred to in the CDM Regulations. See www.hse.gov.uk

107 This shows show the UK is given freedom to interpret and implement EU Directives. However there is a difference between UK practice and practice on mainland Europe. A case in Italy, the Nussbaumer case involving the replacement of a roof to a house in Merano, Bolzano in Italy where Mrs Nussbaumer, the owner, was the H&S supervisor may change the way H&S regulations apply to projects for homeowners.

108 Office of National Statistics Economic Review April 2012; www.ons.gov.uk .

109 The 'Lump' was a major problem for governments as the workforce was completely unregulated and were paid in cash thus avoiding paying income tax and national insurance. It suited companies too as these casual employees did not have any employment rights. A similar situation prevailed in many industries including the docks which relied on a mobile workforce. Over time greater unionisation, especially in the docks, led to improvements in employment, but the construction industry remained as a poor employer – paying badly (if in cash) with little investment in training or welfare. This also contributed to the poor health and safety record of construction sites.

110 'Constructing the Team' (The Latham Report) Stationery Office 1996.

111 'Constructing Excellence' (The Egan Report) Stationary Office 1998.

112 www.constructingexcellence.org.uk Home page accessed 22.9.12.

113 www.constructingexcellence.org.uk Key Performance Indicators accessed 22.9.12.

114 See Murdoch J & Hughes W 'Construction Contracts' (4th ed.) p. 41.

115 Murdoch J & Hughes W p. 55.

116 www.kone.com accessed 26.6.12.

117 www.wilmottdixongroup.com accessed 4.7.12.

118 JCT 'Constructing Excellence' contract Section 2 clause 2.1.

119 www.jctcontracts.com The JCT website is also a valuable source of information about different construction contracts. The examples given in the text are generic as the contracts are updated regularly. It is important to use the most up-to-date contract.

120 www.neccontract.com

121 www.fidic.org. FIDIC, the International Federation of Consulting Engineers publish a suite of contracts used on civil engineering and construction projects throughout the world.

122 Finch R 'NBS Guide to Tendering: for construction projects' RIBA 2011.

123 For example, for further information see JCT Practice Notes 6 www.jctcontracts.com

124 The Consolidated Directive on Public Procurement (2004/18/C) For further information on the procurement rules see www.out-law.com/page-5964 accessed 10.7.12.

INDEX